Green Intelligence

Carbon Footprint Assessment in Digital Economy

Editors

Larisa Ivascu

Politehnica University of Timisoara
Faculty of Management in Production and Transportation
Timisoara, Romania

Marius Pislaru

"Gheorghe Asachi" Technical University of Iaşi
Faculty of Industrial Design and Business Management
Iaşi, Romania

Muddassar Sarfraz

School of Management
Zhejiang Shuren University
Hangzhou, Zhejiang, PR China

CRC Press
Taylor & Francis Group
Boca Raton London New York

CRC Press is an imprint of the
Taylor & Francis Group, an **informa** business
A SCIENCE PUBLISHERS BOOK

First edition published 2026
by CRC Press
2385 NW Executive Center Drive, Suite 320, Boca Raton FL 33431

and by CRC Press
4 Park Square, Milton Park, Abingdon, Oxon, OX14 4RN

CRC Press is an imprint of Taylor & Francis Group, LLC

Library of Congress Cataloging-in-Publication Data (applied for)

ISBN: 978-1-032-84012-3 (hbk)
ISBN: 978-1-032-85612-4 (pbk)
ISBN: 978-1-003-51899-0 (ebk)

DOI: 10.1201/9781003518990

Typeset in Times New Roman
by Prime Publishing Services

Preface

In an era of rapid digital transformation, assessing and reducing carbon footprints has become critical to achieve sustainable development. This book investigates how digital technologies, intelligent systems, and data analytics are reshaping environmental accountability and green practices. It delves into strategies for integrating sustainability into digital economies, highlighting real-world applications, case studies, and policy implications. Readers will gain insights into the role of digital innovation in driving low-carbon transitions, fostering eco-efficiency, and enabling climate-resilient growth. Bridging theory and practice, this volume serves as a valuable guide for academics, professionals, and decision-makers committed to a greener, smarter future.

"Green Intelligence: Carbon Footprint Assessment in the Digital Economy" is an essential topical volume, as it combines sustainability with digital transformation. Sustainability is a concept increasingly addressed by many organizations. The digital economy and green intelligence complement concerns for the environment and the health of society.

"Green Intelligence" refers to the use of technology and smart data to reduce the impact on the environment, especially in the context of economic and digital activities. It is a combination of several important concepts for researchers and professionals in digital technology (AI, IoT, Big Data), sustainability and smart decision-making processes, based on environmental assessments.

Carbon footprint assessment involves measuring the amount of greenhouse gas emissions (CO_2 and equivalents) generated by an organization, product or service. In the digital economy, this involves data on emissions generated by data centers, energy consumed by devices and networks, and indirect emissions from the digital supply chain.

The digital economy (cloud computing, AI, blockchain, etc.) is highly energy dependent. Data centers consume huge amounts of electricity, often from unsustainable sources.

Carbon footprinting helps identify major sources of emissions, optimize infrastructure, generate energy using renewable sources, improve processes, and reduce defects.

Green Intelligence and carbon footprinting are essential for a sustainable digital economy. They not only reduce environmental impact, but also help companies become energy efficient, compliant with legislation, and more attractive to customers and investors.

This volume addresses concepts about the behavior of universities and their digital transformation, the role of digitalization in organizations, strategies for improving capacity for the circular economy, neuro-fuzzy modeling, information models for improving footprint, green intelligence through AI and STEM skills, green energy management in organizations, innovative solutions for the fashion and pharmaceutical industries, and green Lean Six Sigma. The complete and comprehensive approach to the concepts makes this volume a valuable tool for the field of sustainability.

This volume is an important support for professionals, researchers, teachers, students, doctoral students and society. Each concept is comprehensively evaluated, contributing to the advancement of research conducted to date.

Editors

Contents

Chapter 1

Sustainability in the Digital Age Digital Transformation and University Rankings for Future Leaders

Larisa Ivascu,[1,*] *Alin Emanuel Artene*[1,*] and *Aura Emanuela Domil*[2]

1. Introduction

For decades universities set out to be greener, save energy, reduce waste and lead by example and operated in the dark, making decisions without enough data and without the tools to fully understand their environmental footprint.

Universities are not only places where knowledge is passed down but also hubs of innovation with the resources, brainpower, and trust of the public, making them the perfect institutions to pioneer green intelligence (Mohamed Hashim et al., 2022). Within university campuses we can create a model of sustainability that can be scaled, replicated, and improved globally because every student that has ever been a student at a sustainable university becomes an ambassador for sustainability, carrying the lessons and culture into their careers, home and into the world.

University environment is uniquely positioned to design strategies for implementing sustainable development goals and guidelines because it reflects more an environment of learning but hub of innovation, collaboration, and societal transformation. These aspects make the university environment a perfect one for tackling complex, multidimensional challenges like sustainability. Because higher education institutions are trusted, they can serve as neutral conveners of dialogue,

[1] Politehnica University of Timisoara, Faculty of Management in Production and Transportation, 14 Remus Street, Timisoara, Romania.
[2] Faculty of Economics and Business Administration, West University of Timisoara, 16 Pestalozzi Street, 300115 Timisoara, Romania.
* Corresponding author: larisa.ivascu@upt.ro, alin.artene@upt.ro

DOI: 10.1201/9781003518990-1

hosting think tanks, workshops, and roundtables and creating spaces for companies, municipalities, NGOs, and other stakeholders to collaborate on sustainability goals (Mohamed Hashim et al., 2022). These partnerships are essential for addressing challenges that no single organization or sector can solve alone and helping companies understand how operations affect local communities or guide a municipality in designing policies that align with global ESG standards while meeting local needs.

In our opinion, universities have the skills and knowledge to advise companies on reporting progress in sustainability while also addressing the broader ethical aspects of the United Nations Sustainable Development Goals. This is because universities are uniquely equipped to guide organizations in navigating complexities as they combine deep expertise with a commitment to societal progress (Ikenwe and Udem, 2022). One of the challenges companies face today is how to effectively measure, track, and communicate their progress towards sustainability goals because Sustainable Development Goals (SDG) reports require not only technical expertise but also an understanding of transparency, accountability, and stakeholder engagement. Universities, with their vast experience in research and data analysis, are well-positioned to support this process, guiding companies in collecting high-quality data, establishing baselines, and setting measurable targets (Hashim et al., 2022). They can also guide companies in managing these questions by fostering dialogues that bring together diverse perspectives. In addition, universities play a critical role in educating future leaders who understand the intersection of sustainability and ethics by offering interdisciplinary programs, experiential learning opportunities and preparing students to think critically about the SDG.

Sustainability is not just about environmental responsibility but also about creating systems that are equitable, resilient, and accessible to all university students and staff. Digital services must minimize their environmental impact by reducing energy consumption in data centers, using sustainable hosting platforms, and optimizing the software to run efficiently on various devices, prioritizing cloud services powered by renewable energy or implement algorithms that minimize data usage and processing power. University leaders must make sure that digital services are accessible to all students, regardless of socioeconomic background, physical ability, or geographic location because accessibility is a fundamental principle of equity, and it is critical that no student is left behind due to poorly designed or exclusive systems. Sustainable digital services should be designed with long-term cost efficiency in mind, that includes reducing redundancy, leveraging open-source tools where possible, and building scalable systems that can adapt to changing demands (Trevisan et al., 2024).

In a world that is increasingly digital, universities must lead by example in demonstrating how to design systems that are both sustainable and accessible, by embedding these principles into their digital infrastructure and reducing their environmental impact, creating systems that endure, adapt, and serve, not just the students of today, but the generations to come (Carayannis and Morawska-Jancelewicz, 2022). University rankings provide a comprehensive picture

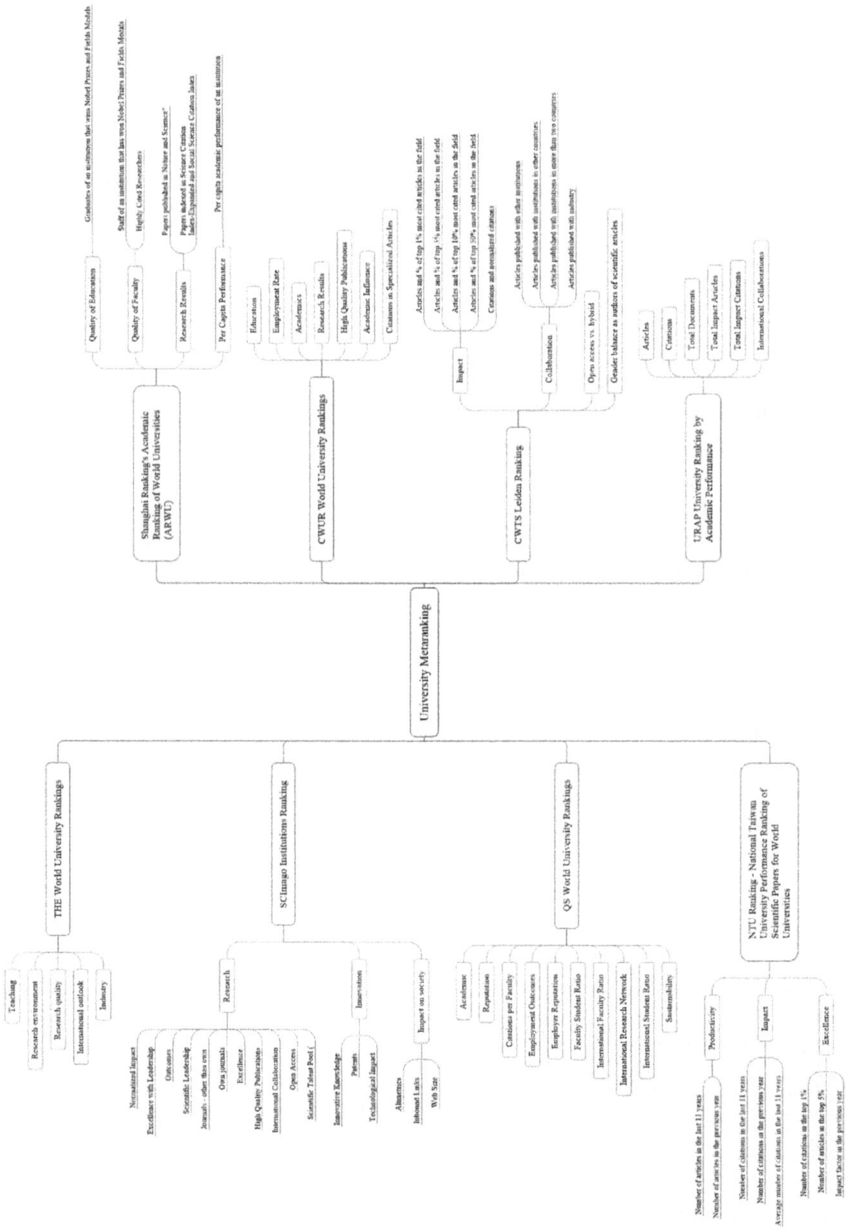

Figure 1(a). A comprehensive analysis of the rankings included in UM.

of the potential of universities. At a global level, there are many rankings that list universities. They use a series of indicators that they establish according to their own methodology, often less transparent. At a national level, in Romania, in recent years the National Metaranking has been carried out. During the period 2017–2021, the University Metaranking (UM) was published, and in 2021, the process for the National Metaranking (NM) was initiated.

The purpose of this UM is to identify the performance of universities, to establish a hierarchy of institutions and to provide the decision-making process with a complex analysis of the directions of university activity. This UM includes the results of eight international rankings: Shanghai Ranking's Academic Ranking of World Universities (ARWU), CWUR World University Rankings, CWTS Leiden Ranking, NTU Ranking - National Taiwan University Performance Ranking of Scientific Papers for World Universities, QS World University Rankings, SCImago Institutions Ranking, THE World University Rankings, and URAP University Ranking by Academic Performance.

A comprehensive analysis of these rankings highlights the main strategic directions and their indicators. These are presented in the following figure.

Analyzing the top seven universities in the QS World University Rankings for Sustainability in 2024, several patterns emerge. These highlight the common strengths contributing to their high rankings, suggesting that a multifaceted approach, combining strong research output, educational integration, community engagement, and institutional commitment to environmental, social, and governance (ESG) principles are instrumental in achieving top positions in sustainability rankings as shown in Figure 1(b).

These top universities excel in sustainability because they embody a powerful synergy of diverse regional representation, cutting-edge research excellence, and robust ESG-driven strategies. They seamlessly integrate sustainability into their educational curricula, equipping future leaders with the skills to address global challenges, while fostering impactful community engagement that bridges the local and the global needs. Their commitment to transparency, ethical governance, and long-term social responsibility demonstrates that true sustainability is not just about

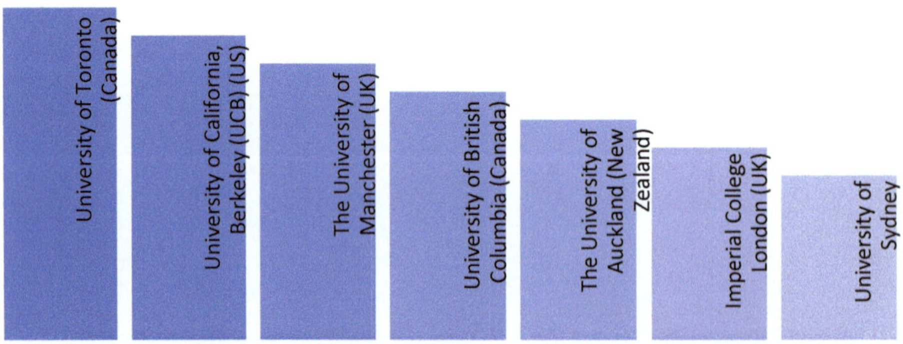

Figure 1(b). Top 7 QS World University Rankings for Sustainability in 2024.

Figure 2. Key patterns of sustainability excellence in top universities.

solving today's problems but transforming systems and shaping a resilient, equitable future for all, Figure 2.

2. Drivers and Barriers to sustainability in higher education campus

When we think about the drivers of university sustainability, we tend to imagine forces that propel universities toward sustainability and at the heart of it lies a moral imperative because universities are custodians of societal progress. But with this privilege comes a responsibility to lead by example, to tackle the grand challenges of our time, and to prepare the next generation not just to navigate a changing world but to improve it. Sustainability is no longer a peripheral concern; it affects every aspect of life. To ignore it would be a betrayal of the very purpose of higher education.

Beyond this ethical dimension, we are driven by the demands of the students we serve, who today are more aware, more engaged, and more insistent than ever. They ask what is the university doing about climate change, how is it addressing inequality, are its actions aligned with the values it espouses? (Carayannis and Morawska-Jancelewicz, 2022). These questions reflect a deep expectation that universities should lead by example. But sustainability is not just about obligation, it also represents an opportunity. Universities are uniquely positioned to innovate because they have the resources, the intellectual capital, and the creative freedom to experiment with bold ideas. By turning campuses into living laboratories for sustainability, universities can pioneer energy-efficient technologies, explore the circular economy, and test new models of urban resilience. In doing so, they do more than reduce their carbon footprint and generate knowledge that benefits society at large. Within this higher education environment, sustainability, when embraced fully, becomes a source of distinction and leadership (George and Wooden, 2023; Hashim et al., 2022). We must also mention the practical drivers. While it may seem counterintuitive, sustainability often makes economic sense on a long term.

Table 1. Drivers of university sustainability.

Drivers	University opportunity	Practical applications in campus
Moral imperative to lead by example	Position universities as ethical leaders addressing global challenges	Develop campus-wide sustainability policies and ethical guidelines for operations
Demand from students and society	Engage with student-led initiatives and meet societal expectations	Incorporate sustainability into curricula and offer mentorship for student sustainability projects
Potential for innovation and leadership	Develop new technologies and sustainability models that enhance global reputation	Turn campuses into living labs to test renewable energy systems or smart infrastructure
Financial efficiency and long-term gains	Lower operational costs while advancing environmental goals	Implement energy-efficient building designs and waste reduction programs
Alignment with global frameworks (SDGs, ESG)	Strengthen global partnerships and enhance rankings through alignment with sustainability goals	Publish sustainability reports aligned with SDG/ESG frameworks and collaborate internationally
Increasing climate-related risks	Mitigate climate risks and enhance campus resilience through proactive strategies	Install renewable energy sources like solar panels to minimize climate-related disruptions
Rising pressure from accreditation bodies	Demonstrate compliance and innovation to meet accreditation standards	Embed sustainability criteria in accreditation submissions and strategic plans
Technological advancements enabling sustainability	Utilize digital tools and IoT to improve campus sustainability management	Deploy IoT sensors to monitor energy, water, and waste usage in real time
Global competition for reputation and rankings	Attract global students and researchers by showcasing sustainability excellence	Host international sustainability conferences and attract environmentally conscious students
Opportunities to build cross-sector partnerships	Collaborate with industries and governments for shared sustainability goals	Develop public-private partnerships for green infrastructure projects
Societal expectations for ethical responsibility	Incorporate ethical frameworks into research, operations, and governance	Offer training programs on ethics and sustainability for faculty, staff, and students

These benefits come with short term costs like retrofitting buildings, installing renewable energy systems and developing new curricula, but the long-term savings are undeniable. Energy-efficient buildings, for example, reduce utility costs, while waste management systems create efficiencies that save money (Weiss et al., 2021). The drivers for university sustainability are presented in Table 1.

Despite strong motivations propelling universities toward sustainability, numerous obstacles persist, the first of them being, financial limitations. For many institutions, particularly those in less affluent regions, sustainability initiatives can feel like a distant ambition. Here, the focus often remains on meeting basic operational needs. The stark reality for some universities is not deliberating over how to create a green campus but struggling with questions as fundamental as keeping essential services running. Limited budgets, coupled with competing institutional priorities, often relegate sustainability to the status of an optional, rather than essential pursuit. Even for well-funded institutions, there's another significant hurdle to overcome in the form of the resistance to change. Universities are paradoxical entities, centers of innovation on one hand, and protectors of long-standing traditions on the other. Efforts to introduce sustainability programs frequently encounter entrenched practices, fragmented departmental operations, and administrative inertia. This resistance is often subtle rather than outspoken—a quiet hesitancy, a collective feeling that sustainability is someone else's responsibility, or simply a sense of comfort with maintaining the existing systems (Weiss et al., 2021).

Another formidable challenge lies in infrastructure because data is the backbone of sustainability when it comes to understanding energy consumption, tracking emissions, and evaluating waste. Without the right technological tools, universities are effectively navigating blindfolded, unable to fully grasp or manage their environmental impact. This problem is particularly acute in institutions where outdated infrastructure dominates, necessitating not only costly upgrades to systems but also substantial investments in training personnel to handle new tools and methodologies.

Perhaps the most complex issue is how to measure sustainability's impact. The interconnected nature of sustainability initiatives, combined with the long timelines often needed to see their effects, makes this a particularly daunting task (Akins et al., 2019). Quantifying the benefits of a newly adopted energy policy, for instance, is no straightforward endeavor. Similarly, determining the success of sustainability-focused curricula may take decades, as the true impact often becomes evident only after graduates enter the workforce. Without widely accepted standards for measurement, it becomes challenging for universities to gauge their progress, defend their spending, or effectively communicate their achievements to stakeholders. The lack of standardized metrics creates uncertainty, and uncertainty breeds inaction (Blanco-Portela et al., 2018).

All these challenges, though significant, are by no means insurmountable. Each one presents an opportunity for universities to innovate, adapt, and take deliberate action in addressing the structural and cultural changes required to lead in sustainability, Table 2.

The challenge of meeting the European Union's 2030 environmental targets is one that demands collaboration, innovation, and determination across all sectors and the role of higher education universities is extremely important (Carayannis and Morawska-Jancelewicz, 2022). The European Environment Agency stated that the road ahead will be challenging and the universities can think critically, adapt, and lead, being not just centers for learning, but symbols of progress, spaces where ideas transcend borders and where the pursuit of knowledge is intrinsically tied to the betterment of society. In this, sustainability is not a choice but an obligation. University campuses have a duty to reflect the values of the future they seek to create. It is this ethos of leadership, of ethical stewardship, that aligns seamlessly with the EU's vision for a greener, more sustainable Europe. Equally compelling is the voice of students that populate the campuses demanding action, not just rhetoric,

Table 2. Threats for university sustainability.

Barriers	**University threats**	**Practical concerns in campus**
Financial constraints limiting sustainability investments	Missed opportunities for long-term cost savings and operational efficiency	Identify and secure external funding sources to address financial gaps
Institutional resistance to change	Stagnation due to adherence to outdated practices and traditions	Develop change management strategies to foster cultural shifts
Lack of technological infrastructure	Inability to track, analyze, and optimize sustainability efforts effectively	Invest in IT infrastructure and data management systems to support sustainability tracking
Difficulty in measuring sustainability impact	Failure to demonstrate progress leads to reputational damage	Establish standardized metrics for evaluating sustainability initiatives
Competing priorities within universities	Sustainability is deprioritized, leading to delays in implementation	Create dedicated sustainability offices to prioritize and coordinate efforts
Limited expertise in sustainability and digital transformation	Over-reliance on external consultants or insufficient internal capacity	Offer professional development programs on sustainability and digital tools
Fragmentation across university departments	Ineffective collaboration and duplication of efforts across departments	Implement cross-departmental task forces to improve collaboration
Challenges in integrating sustainability into curricula	Students are not equipped with skills to address future sustainability challenges	Develop modular and flexible sustainability courses to integrate into curricula
Insufficient funding for research and innovation	Reduced competitiveness in securing research funding and partnerships	Seek partnerships with industries and governments to co-fund research
Regulatory and compliance complexities	Difficulty maintaining compliance with evolving global standards	Streamline compliance processes and leverage technology for regulatory reporting
Geographic and demographic disparities	Unequal access to sustainability resources across campuses and regions	Develop tailored solutions for regional campuses to address unique challenges

from the institutions that educate them. They expect universities to model sustainable behavior, to integrate it into every facet of campus life. This demand is a profound driver of change because it is immediate not abstract. A university campus has the extraordinary potential for innovation including climate-resilient technologies to cutting-edge renewable energy solutions, pioneering the tools that will enable the EU to meet its environmental objectives. By integrating the European Green Deal and SDG principles into their strategies, universities and their virtual and physical campuses position themselves as global leaders, capable of driving policies and partnerships that extend far beyond their campuses (Blanco-Portela et al., 2018).

Resistance to change also casts a long shadow because universities, paradoxically, are both agents of progress and guardians of tradition. Bureaucracy, entrenched

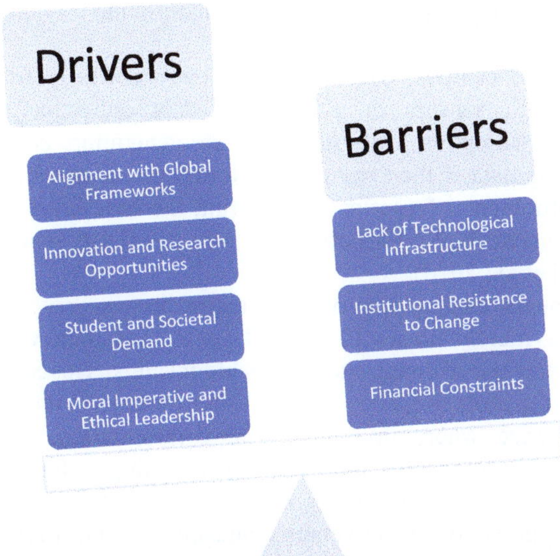

Figure 3. Drivers and barriers in higher education supporting EU environmental targets.

practices, and silos can stymie even the most well-intentioned sustainability initiatives. Overcoming this inertia requires cultural shifts within institutions representing a shift that demands time, effort, and strong leadership. A further challenge is the lack of technological infrastructure within university campuses. Sustainability in the digital age relies heavily on data and on the ability to monitor emissions, measure energy usage, and track progress, Figure 3. For universities that lack this infrastructure, it is akin to navigating these challenges (Akins et al., 2019).

Unfortunately for university staff, sustainability is complex, and it unfolds over decades and encompasses interconnected systems and university managers. The shareholders should quantify the ripple effects of a new energy policy, not to mention the success of a curriculum that teaches sustainability, when its true outcomes may only manifest decades later (Leal Filho et al., 2019).

All the opportunities and barriers presented above leave universities with a dual challenge. On one hand, they must harness the drivers fueled by ethics, innovation, societal demand, and alignment with global goals and on the other hand, they must systematically dismantle the barriers that hold them back: the financial gaps, the institutional inertia, the lack of infrastructure, and the complexity of measurement.

3. Green intelligence framework for sustainable campus management

Green intelligence framework for sustainable campus management cannot function as a space of education and must be reshaped as an ecosystem of sustainability and intelligence. This vision centers on the concept of green intelligence, a structured, multi-layered approach to creating sustainable campus management systems grounded in practical application, designed to transform the way universities operate and their impact on the environment.

In a sustainable campus, data needs to flow seamlessly from every corner communicating with sensors in a real-time data collection and integration, tracking energy consumption, water usage, waste generation, and even patterns of transportation. This data is then aggregated, forming a comprehensive map of the university's environmental footprint and offering the base layer of the framework in the shape of a network of information that provides clarity where there was once opacity (Agarwal et al., 2020).

Future campuses cannot rely on data alone because raw numbers, without interpretation, are merely noise. The next layer of our framework is where this data is analyzed and transformed into actionable insights. This second layer equips universities with the tools to go beyond reacting to problems and takes the form of a layer of intelligence, where algorithms process information, detect patterns, and predict outcomes. AI will identify inefficiencies in heating systems, forecasting peak electricity demand, or even anticipating maintenance needs before equipment fails. The processing layer empowers university management teams to predict and prevent future issues (Mendoza et al., 2019).

Once insights are derived, the abstract becomes tangible, the framework moves to the decision-making layer which is the output layer. Data is translated into dashboards, visualized in ways that are intuitive and accessible to campus leaders. The sustainability office may have access to viewing real-time energy metrics on their screen, identifying a spike in consumption, and receiving a system-generated recommendation to optimize HVAC systems in underutilized spaces. This output layer is about clarity and making complex systems understandable, so that action can be swift and impactful.

The mission, vision and objectives of the sustainability department of the university must lead to a campus that operates as a dynamic, intelligent organism and the action taken must lead to implementation. This is the post layer where the framework touches the physical world and where recommendations are enacted through a combination of automation and human intervention. For example, smart lighting systems can adjust based on occupancy or automated waste-sorting systems can reduce landfill contributions. For other challenges that require human expertise, the system flags the issues and empowers facilities teams to respond quickly and effectively. The final step in this multi-layer process, the one that ensures this framework is not static, but ever-evolving as a continuous improvement phase because sustainability is a journey that requires adaptation and learning and not a destination. This layer evaluates the impact of actions taken, measures progress against benchmarks and identifies opportunities for refinement representing a space where feedback loops are closed, where new data feeds back into the system to enhance its predictive power and effectiveness (Zhang, 2023).

Because the proposed sustainable digital campus model is not just a technical solution, it must be embraced by every stakeholder on campus, from students to faculty, from administrative staff to facilities teams, everyone must see themselves as part of this ecosystem. A sustainable university framework is not the job of a single

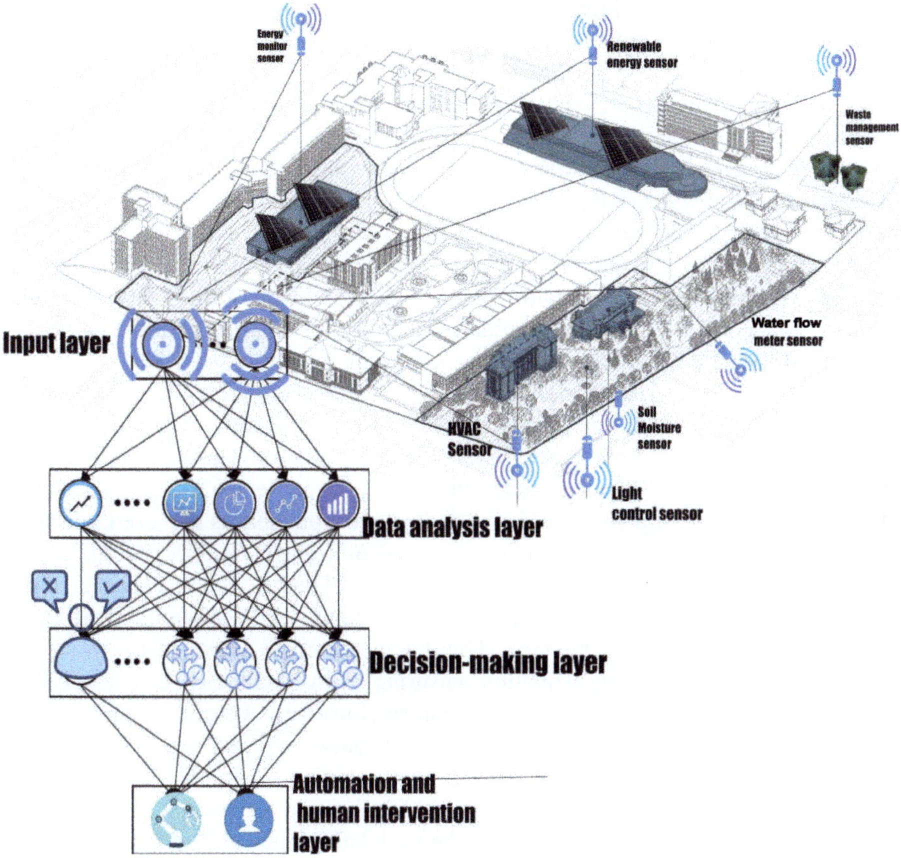

Figure 4. Politehnica University of Timisoara (UPT) - green sustainable intelligent campus model.

office or department but a collective responsibility. The added value of the proposed framework presented in Figure 4 is the scalability and adaptability of a system that learns, grows, and evolves with the institution it serves.

We want to emphasize the importance of a university campus not as a collection of disconnected buildings and pathways but as a living, breathing organism that requires a nervous system to sense, interpret, and respond to its environment ("Roles of Education in Creating Awareness of Sustainability Campus Management for Green Campus Concept," 2023). This is precisely the role that sensors play in creating a sustainable campus because they are the eyes and ears of a system that continuously learns and adapts, ensuring that resources are used wisely, emissions are minimized, and the well-being of everyone within the campus is safeguarded.

In Figure 4 which depicts the Politehnica Timisoara University digital sustainable campus model, the sensors must be integrated as invaluable tools in the journey toward sustainability. Energy monitoring sensors placed across the campus buildings, provide real-time data on electricity consumption and offer input on how and when energy is used, identifying patterns, and making decisions that align with both cost-effectiveness and environmental stewardship.

By analyzing this data, UPT management will be able to optimize energy usage, schedule maintenance to prevent waste, and even adjust building operations to align with fluctuating energy demand.

Water flow meters are another vital component of the above campus framework's sustainability strategy because these devices allow the university to track water usage with precision, detect leaks almost instantly, and make informed decisions about conservation. Environmental sensors play a pivotal role in ensuring the campus is not just sustainable but also a healthy place to learn and work, that is why air quality monitors placed in lecture halls, laboratories, and common areas help the university understand how indoor conditions change over time.

When CO_2 levels rise or air circulation becomes insufficient, the system alerts the user, enabling it to act quickly, whether that means adjusting ventilation systems or simply encouraging occupants to take a break in fresh air. This attention to detail ensures that sustainability does not come at the expense of human experience. Occupancy sensors help the university understand how spaces are used, ensuring that lighting and HVAC systems operate only when needed (Gandasari et al., 2020).

In the rector's management plan, the campus will be full of green spaces, places for students to relax, reflect, and connect with nature and maintaining the green spaces sustainably requires balance. This balance can be achieved by using soil moisture sensors, optimizing irrigation, ensuring plants receive exactly the amount of water they need, the so-called precision agriculture but on a micro scale, serving both the aesthetic and environmental goals of the university campus.

These sensors, individually small and seemingly simple, become truly transformative when they work together within a layered framework represented in Figure 4 through the neural network. At the foundation, they collect data, providing raw information that is essential for understanding the campus ecosystem. Then, this data flows upward, into systems that analyze and interpret it, identifying inefficiencies, predicting future needs, and recommending actions.

At the next level, this intelligence is visualized through dashboards and reports, making it accessible and actionable for decision-makers. And finally, the framework enables action, whether through automated systems that adjust settings in real time or through human intervention guided by these insights.

On the UPT campus, this layered approach will contribute to a sustainable future, not only reducing resource consumption but also fostering a culture of sustainability. Students, faculty, and staff hopefully will see the impact of these systems in their daily lives, whether it's in the improved air quality, more vibrant green spaces, or reduced energy costs that free up resources for educational programs (Mejia et al., 2019).

In Table 3, you can see that the sensors are catalysts for change, enabling the university to reimagine what a university can be in the 21st century, a place where students are eager to lead, the policymakers are eager to shape the collective future, the vision is achievable, and the impact is profound (Fu et al., 2022). The evolving paradigm of AI and digital connectivity also presents an opportunity for

Table 3. Sensors for Sustainable campus management at UPT.

Sensors	Application	Use cases for UPT campus
Energy Monitoring Sensors	Devices that track electricity consumption across campus facilities, identifying areas for energy optimization.	Optimizing energy usage across buildings to reduce costs and emissions.
Environmental Sensors	Instruments measuring indoor air quality, temperature, humidity, and CO_2 levels to ensure healthy learning environments.	Ensuring optimal indoor air quality in lecture halls and laboratories.
Occupancy Sensors	Sensors detecting room usage patterns, aiding in space utilization and energy conservation by adjusting lighting and HVAC systems based on occupancy.	Improving space utilization and reducing energy waste in unused rooms.
Water Flow Meters	Devices monitoring water usage to detect leaks and promote water conservation strategies.	Detecting water leaks promptly to prevent wastage and promote conservation.
Waste Management Sensors	Sensors in waste bins that monitor fill levels, optimizing waste collection routes and promoting recycling efforts.	Streamlining waste collection efforts and promoting better recycling practices.
Renewable Energy Generation Monitors	Systems tracking the performance of on-campus solar panels or wind turbines, ensuring optimal energy production.	Monitoring the efficiency of renewable energy systems to maximize their contribution.
HVAC System Sensors	Devices monitoring heating, ventilation, and air conditioning systems to maintain efficiency and indoor comfort.	Maintaining optimal indoor temperatures and improving energy efficiency.
Lighting Control Sensors	Smart lighting systems that adjust based on natural light availability and occupancy, reducing energy consumption.	Reducing unnecessary energy consumption by leveraging natural light and managing occupancy.
Soil Moisture Sensors	Instruments used in campus landscaping to monitor soil moisture levels, optimizing irrigation and conserving water.	Conserving water while maintaining the greenery and aesthetics of the campus.
Transportation and Parking Sensors	Devices tracking campus shuttle usage and parking space occupancy to improve transportation planning and reduce emissions.	Optimizing shuttle schedules and parking usage to minimize emissions and improve efficiency.

universities to amplify their impact. AI can enhance the university experience by providing personalized learning pathways, automating administrative tasks, and analyzing data to improve decision-making. However, these technologies are tools, not replacements. Their effectiveness depends on the human ingenuity and ethical frameworks provided by academic institutions. It is universities that ensure technology serves humanity, not the other way around.

4. Conclusions

In conclusion, we can reflect on the challenges we face, the opportunities we hold, and the responsibilities we must embrace as institutions of higher learning. The theoretical model that has taken shape in this chapter is rooted in the profound belief that universities are more than educational establishments, being engines of progress, arbiters of ethical leadership, and the incubators of solutions to the world's most pressing problems. A university is not just a place of learning, but a crucible of culture, a community where values are shaped, habits of thought and action are formed, and the foundations of a sustainable mindset are laid. Universities, by cultivating sustainable practices within their own operations and embedding sustainability into their curricula, serve as living examples of the change they seek to promote (Yeo, 2018).

Throughout this exploration, we have examined the drivers that propel universities toward sustainability. The moral imperative, the expectations of students and society, the boundless potential for innovation, and the alignment with global frameworks like the European Green Deal. Universities are uniquely positioned to lead, not only because they possess the intellectual capital to innovate but because they embody the values and aspirations of future generations. Every decision made on a university campus ripples outward, influencing students who will go on to shape policy, industries, and communities. In this way, universities hold a profound responsibility to lead by example, modeling the sustainable practices we wish to see in the broader world.

In our endeavor to identify the framework upon university sustainability in the digital age, we identified barriers that cannot be ignored like financial constraints, resistance to change, and the lack of technological infrastructure presenting real and pressing challenges. These barriers, though formidable, are not insurmountable and they are calls to action, urging us to rethink traditional approaches and seek innovative solutions. And while the complexity of measuring impact remains an ongoing obstacle, it also serves as a reminder of the intricate interconnectedness of sustainability efforts in the form of a challenge that demands our collective ingenuity and perseverance.

Our approach emphasized the pivotal role of technology in this endeavor through the sensors presented and devices explored that are not just tools, but, the silent architects of a new kind of campus that learns, adapts, and grows. These technologies enable universities to measure their impact in real time, to manage resources with precision, and to create environments that are as sustainable as they are conducive to learning. But technology, as we have acknowledged, is only one part of the equation. Its true power lies in its integration within a broader framework like a Green Intelligence Framework that connects data to action, decision-making to impact, and innovation to culture.

The principles we have discussed in this chapter, like sustainability, innovation, and the integration of technology, are not confined to the pages of a book or the boundaries of a campus. They are principles that speak to the very essence of what it means to be a university in the XIst century, challenging us to redefine our roles, to extend our influence beyond traditional academic pursuits, and to become active participants in the global effort to create a sustainable future.

World leaders and prominent thinkers have also emphasized the pivotal role of higher education institutions in shaping future leaders, underscoring the necessity for universities to adapt their curricula to meet the evolving demands of the workforce, particularly in the context of the Fourth Industrial Revolution. They advocate for educational models that foster collaboration, system leadership, and agility among students, preparing them to navigate and lead in a rapidly changing world. Furthermore, UNESCO's report on leadership in higher education discusses the evolving requirements of effective leadership within academic institutions. It highlights the importance of universities in mobilizing various actors and resources to address global challenges, thereby reinforcing their role in cultivating the next generation of leaders equipped to tackle complex societal issues.

In an era marked by escalating social and political unrest, coupled with the undeniable impact of increasingly frequent natural disasters, the role of universities as beacons of progress and responsibility has never been more critical. The expectation for universities to become sustainable in the digital age is imperative because our world is facing profound challenges that transcend national borders. Climate change is no longer a distant threat but a present reality, with wildfires, floods, and hurricanes reshaping landscapes and livelihoods. At the same time, societal fractures, marked by inequality, polarization, and the erosion of trust in institutions, demand solutions that are as inclusive as they are innovative. Amidst these crises, universities stand uniquely positioned as spaces where knowledge, dialogue, and innovation converge. They are among the few institutions capable of addressing these multifaceted challenges holistically and, sustainability, in this sense, must go beyond operational efficiency or green buildings encompassing a broader role that integrates environmental responsibility, social equity, and digital innovation into every facet of university life. Universities must not only reduce their own carbon footprints but also lead by example, demonstrating how complex systems can adapt to the pressing demands of sustainability becoming living laboratories for the future, places where students, researchers, and communities come together to test and implement solutions that extend far beyond campus boundaries.

While it is true that the internet offers unprecedented access to knowledge and that AI can process and deliver information with remarkable speed, universities are not, and have never been, mere repositories of facts. Their value lies in their ability to transform information into understanding, to foster critical thinking, and to nurture wisdom. In a world awash with data, it is universities that teach us how to discern, evaluate, and apply that information in meaningful ways. AI can analyze, simulate and predict, but it cannot teach us empathy, moral judgment, or the art of collaboration. These are human skills or leadership skills, that universities are uniquely positioned to cultivate. Universities provide something that the digital world cannot: because they are spaces where knowledge is not consumed in isolation but explored in dialogue and they bring together diverse perspectives, disciplines, and experiences, creating a fertile ground for innovation and ethical reasoning.

References

Agarwal, P., Ravi Kumar, G. V. V. and Agarwal, P. (2020). IoT based framework for smart campus: COVID-19 readiness. Proceedings of the World Conference on Smart Trends in Systems, Security and Sustainability, WS4 2020. https://doi.org/10.1109/WorldS450073.2020.9210382.

Akins, E. E., Giddens, E., Glassmeyer, D., Gruss, A., Hedden, M. K., Slinger-Friedman, V. et al. (2019). Sustainability education and organizational change: A critical case study of barriers and change drivers at a higher education institution. Sustainability (Switzerland) 11(2). https://doi.org/10.3390/su11020501.

Blanco-Portela, N., R-Pertierra, L., Benayas, J. and Lozano, R. (2018). Sustainability leaders' perceptions on the drivers for and the barriers to the integration of sustainability in Latin American Higher Education Institutions. Sustainability (Switzerland) 10(8). https://doi.org/10.3390/su10082954.

Carayannis, E. G. and Morawska-Jancelewicz, J. (2022). The futures of Europe: Society 5.0 and Industry 5.0 as driving forces of future universities. Journal of the Knowledge Economy 13(4). https://doi.org/10.1007/s13132-021-00854-2.

Fu, Y., Li, Q. and Chang, Y. (2022). Operating mechanism of university innovation and entrepreneurship education and its contribution to regional economy. International Journal of Emerging Technologies in Learning 17(14). https://doi.org/10.3991/ijet.v17i14.32917.

Gandasari, I., Hotimah, O. and Miyarsah, M. (2020). Green campus as a concept in creating sustainable campuses. KnE Social Sciences. https://doi.org/10.18502/kss.v4i14.7853.

George, B. and Wooden, O. (2023). Managing the strategic transformation of higher education through artificial intelligence. Administrative Sciences 13(9). https://doi.org/10.3390/admsci13090196.

Hashim, M. A. M., Tlemsani, I., Matthews, R., Mason-Jones, R. and Ndrecaj, V. (2022). Emergent strategy in higher education: postmodern digital and the future? Administrative Sciences 12(4). https://doi.org/10.3390/admsci12040196.

Ikenwe, I. J. and Udem, O. K. (2022). Innovative digital transformation for dynamic information service sustainability in university libraries in Nigeria. Folia Toruniensia 22. https://doi.org/10.12775/ft.2022.004.

Leal Filho, W., Skouloudis, A., Brandli, L. L., Salvia, A. L., Avila, L. V. and Rayman-Bacchus, L. (2019). Sustainability and procurement practices in higher education institutions: Barriers and drivers. Journal of Cleaner Production 231. https://doi.org/10.1016/j.jclepro.2019.05.202.

Mejia, S. P., Hincapie, J. M. M. and Giraldo, J. A. T. (2019). A hub-based university innovation model. Journal of Technology Management and Innovation 14(1). https://doi.org/10.4067/S0718-27242019000100011.

Mendoza, J. M. F., Gallego-Schmid, A. and Azapagic, A. (2019). A methodological framework for the implementation of circular economy thinking in higher education institutions: Towards sustainable campus management. Journal of Cleaner Production 226. https://doi.org/10.1016/j.jclepro.2019.04.060.

Mohamed Hashim, M. A., Tlemsani, I. and Duncan Matthews, R. (2022). A sustainable University: Digital transformation and beyond. Education and Information Technologies 27. https://doi.org/10.1007/s10639-022-10968-y.

Roles of Education in Creating Awareness of Sustainability Campus Management for Green Campus Concept. (2023). International Journal of Advanced Research in Education and Society. https://doi.org/10.55057/ijares.2023.5.4.25.

Trevisan, L. V., Eustachio, J. H. P. P., Dias, B. G., Filho, W. L. and Pedrozo, E. Á. (2024). Digital transformation towards sustainability in higher education: state-of-the-art and future research insights. In Environment, Development and Sustainability (Vol. 26, Issue 2). https://doi.org/10.1007/s10668-022-02874-7.

Weiss, M., Barth, M., Wiek, A. and von Wehrden, H. (2021). Drivers and barriers of implementing sustainability curricula in higher education - assumptions and evidence. Higher Education Studies 11(2). https://doi.org/10.5539/hes.v11n2p42.

Yeo, B. (2018). Societal impact of university innovation. Management Research Review 41(11). https://doi.org/10.1108/MRR-12-2017-0430.

Zhang, X. (2023). Priorities of sustainable smart campus management. In Approaches to Global Sustainability, Markets, and Governance: Vol. Part F175. https://doi.org/10.1007/978-981-19-4005-7_21.

Chapter 2

Role of Digitalization in Organizational Sustainability

Ionuţ Viorel Herghiligiu,[1,2,*] *Marius Pislaru,*[1]
Stefana Catalina Pohonţu-Dragomir[1] and *Cătalin Ioan Budeanu*[1]

1. Introduction

Sustainability and digitalization represent megatrends shaping and facilitating change (Del Río Castro et al., 2021; Brenner and Hartl, 2021). Even if "sustainability" and "digital technologies" appear as disparate terms (George et al., 2021), they contribute to a paradigm shift in social and ecological systems (Gebhardt, 2017). Likewise, as Osburg and Lohrmann (2017) mentions, these trends alter strategic imperatives that are capable of driving major transformations (Guandalini, 2022).

The UN Sustainable Development Goals (SDGs) (UN, 2023a), have emerged as an unprecedented global compass to address current sustainability challenges. Even though this directive framework exists, the field in question is still developing, complex and uncertain. In this sense, the link between digitalization and sustainability reveals major opportunities for shaping a greener economy and society, substantially facilitating the achievement of the SDGs from the macro to the micro level. Although the literature in this field is extensive, the digital paradigm associated with sustainability generates a "gap" that needs to be researched and clarified (Del Río Castro et al., 2021).

Digitalization is recognized as one of the most promising transformations for advancing sustainability (Gouvea et al., 2018) and addressing the SDGs (Sachs et al., 2019; Seele and Lock, 2017; Walker et al., 2019). However, despite its potential, the digital paradigm remains an emerging field, characterized by insufficient theoretical

[1] "Gheorghe Asachi" Technical University of Iaşi, 29 Prof. D. Mangeron Blvd., 700050, Iaşi, Romania.
[2] Academy of Romanian Scientists, Ilfov 3, 050044 Bucharest, Romania.
* Corresponding author: ionut-viorel.herghiligiu@academic.tuiasi.ro

DOI: 10.1201/9781003518990-2

development and numerous technological and non-technological challenges (Del Río Castro et al., 2021).

Digitalization and sustainability are two representative trends that significantly influence the market and organizations. Each has generated a massive amount of research associated with changing management practices, business models, and society more broadly. The intersection of these trends, however, remains a largely unexplored area of research (Osburg and Lohrmann, 2017; Kiron and Unruh, 2018).

Given that organizational sustainability implies "better operational results, lower capital costs, and improved market price" (Pislaru et al., 2019), quantifying significant determinants becomes a pressing necessity. This desirable result can only be achieved through an assessment based on an innovative approach, supported by the fact that, paradoxically, the challenges associated with sustainability remain unresolved (Kohler et al., 2019; Del Río Castro et al., 2021).

Given the ideas mentioned above, it can be stated that this research is important, and represents a current reflection of international and European directions and programs regarding digitalization and sustainability, such as: 2030 UN Agenda (UN, 2023a), the Skills Agenda for Europe for sustainable competitiveness, social equity and resilience (EC, 2020), the UN Environment Program (UN, 2023b), among others.

The implications of this research are also important for practitioners and regulatory authorities, who face both sustainability and digital transformation processes and strive to asses and exploit the relationship between the two "phenomena" (Guandalini, 2022).

2. Digitalization and sustainability – literature background

2.1 Digital transformation

Digital transformation is a complex phenomenon, deeply embedded in the evolution of digital technologies and their ability to generate fundamental changes in the way organizations and societies operate (Ologeanu-Taddei et al., 2024). Digital transformation goes beyond simple digitalization or automation, implying a strategic adaptation at the structural, operational and cultural levels, aimed at creating new values and opportunities through the use of emerging digital technologies (Vial, 2019; Wessel et al., 2021). This evolving concept plays an essential role in redefining business strategies and configuring modern economies, being marked by the rapid dynamics of technological change and the demands of the global market (Ologeanu-Taddei et al., 2024).

Unlike traditional IT-driven organizational transformations, Hanelt et al. (2021) mention that digital transformation redefines organizational identity and structure, emphasizing digital adaptability and agility.

Digital transformation is often defined as a profound and disruptive change in organizational strategies, triggered by digital technologies (Gökalp and Martinez, 2021a,b; Baiyere et al., 2023; Markus and Rowe, 2023), such as: the Internet of Things (IoT) (Gilchrist, 2016), digital twin (Annunziata and Biller, 2015), cyber-physical systems (Schuh et al., 2017), artificial intelligence (Schuh et al., 2017), social networks and platforms, blockchain (Swan, 2015), and everything-as-a-service, robots and drones, big data analysis (Gökalp et al., 2021a,b). This transformation is not limited to the adoption of technologies, but involves restructuring internal processes, reshaping business models and adapting organizational culture to meet the demands associated to a constantly changing market (Ologeanu-Taddei et al., 2024).

Digital transformation can be approached from several perspectives (Wessel et al., 2021; Del Río Castro et al., 2021; Hanelt et al., 2021; Ologeanu-Taddei et al., 2024):

i. *Technological*: adoption of digital technologies, such as AI for optimizing production or IoT for resource management, is central to digital transformation but poses challenges in data security and ethical use (Wessel et al., 2021);
ii. *Strategic*: digital transformation reshapes business strategies, transitioning from traditional models to digital platforms and global networks. Integrating sustainability into these strategies enhances social and environmental impact, aligning with the Sustainable Development Goals (SDGs) (Del Río Castro et al., 2021);
iii. *Operational*: digital transformation involves optimizing processes and creating more efficient workflows. Automation, predictive analytics, and digital collaboration tools are concrete examples of the impact of technology on daily operations (Ologeanu-Taddei et al., 2024).
iv. *Cultural*: the success of digital transformation depends largely on changing organizational culture. Creating a digital mindset and developing organizational agility are essential aspects for the full integration of digital technologies (Hanelt et al., 2021).

Vial (2019), mentions that digital transformation is "a process by which digital technologies create disruptions that determine strategic responses from organizations, thus modifying their value creation paths, while managing structural changes and organizational barriers that influence the outcomes of this process". This definition highlights the disruptive nature of digital transformation, which forces organizations to reconsider their strategic foundations and invest in new skills and modern infrastructures (Ologeanu-Taddei et al., 2024).

Starting from the fact that digital transformation is defined as "... the process by which a firm uses digital technologies to create appropriate new digital business models" (Verhoef et al., 2021; Nguyen and Thanh Hoai, 2022; Ologeanu-Taddei et al., 2024), based on the literature and the aforementioned, the summarized a range of key elements associated with this concept can be observed in Table 1.

Consequently, it can be stated that digital transformation is a complex and multidimensional process through which organizations fundamentally rethink their strategies, business models, processes and operational structures through the integration and extensive use of digital technologies, in order to improve efficiency, generate added value and respond to contemporary challenges (Vial, 2019; Favoretto et al., 2022). This process is often perceived as a driver associated to organizational

Table 1. Organizational digital transformation – associated subdimensions/aspects (Gökalp și Martinez, 2021a,b; Martínez-Peláez et al., 2023; Unlu et al., 2023; EC – DESI, 2023; Ologeanu-Taddei et al., 2024).

No.	Dimension	Sub-dimension
1	Business digital transformation	▪ social media ▪ big data ▪ cloud ▪ electronic information sharing ▪ artificial intelligence ▪ online commerce
2	Digital abilities	▪ internet use ▪ people with IT degrees ▪ digital skills (average; above average; content creation) ▪ IT organizational specialists ▪ training on increasing digital skills
3	Digital infrastructure	▪ fixed infrastructure usage ▪ mobile infrastructure usage

change, leading companies towards flexible organizational models (Hanelt et al., 2021).

2.2 Corporate sustainability

The corporate sustainability concept, originally developed through the Triple Bottom Line (TBL) perspective, proposes the simultaneous integration of the three fundamental dimensions: economic prosperity, environmental integrity and social equity (Elkington, 1994). In this context, corporate sustainability transcends simple social responsibility practices, becoming a mechanism through which organizations can actively contribute to global sustainable development.

The origin of corporate sustainability is deeply rooted in the literature on corporate social responsibility (CSR) and environmental management. Initially, it was viewed through the lens of companies' ecological and social obligations, but it has gradually transformed into a holistic concept that integrates economic and non-economic objectives (Ologeanu-Taddei et al., 2024). Bansal and Roth (2000) highlighted three main motivations for organizations to adopt sustainable practices: competitive advantage, legitimacy, and ecological responsibility. This approach suggests that sustainability is not just a reaction to external pressures, but also a strategic opportunity for long-term value creation.

Ologeanu-Taddei et al. (2024), argue that corporate sustainability is a complex and multidimensional concept that generates tensions between economic, social, and ecological dimensions. The tensions arise from the interdependencies and trade-offs required between these dimensions, which cannot always be balanced simultaneously. Hahn et al. (2018) emphasize that approaching sustainability through a "win-win paradigm" is often unrealistic, and organizations must manage conflicts between their sustainable and economic objectives.

Literature, regarding corporate sustainability, presents the following dimensions (Birkel and Müller, 2021; Seidel et al., 2017; Bolis et al., 2017; Pislaru et al., 2019; Zheng and Walsham, 2021; Ologeanu-Taddei et al., 2024):

i. *Economic/financial*: economic sustainability focuses on ensuring the long-term financial viability of the organization, by maximizing efficiency and reducing costs, but without sacrificing the impact on the environment or society. Sustainable economic strategies include resource optimization, product and process innovation, and the creation of circular business models (Birkel and Müller, 2021).

ii. *Ecological*: environmental protection is a central pillar of corporate sustainability, including practices such as reducing carbon emissions, waste management, and the use of renewable resources. The concept of Green IT and Green IS, highlighted by Seidel et al. (2017), suggests that the use of digital technologies can significantly contribute to reducing the environmental impact of organizations. At the same time, organizations need to manage the dilemmas associated with the use of technology, such as digital waste and high energy consumption.

iii. *Social*: social equity and responsibility towards the community are essential components of corporate sustainability. This involves practices such as respecting employee rights, promoting diversity and inclusion, and supporting social initiatives. In addition, digital technologies offer opportunities to reduce social inequalities through access to resources and information (Zheng and Walsham, 2021).

Embracing corporate sustainability is essential not only for individual organizations, but also for achieving the SDGs established by the United Nations. In a world affected by climate change and social inequalities, companies have a growing responsibility to contribute to solving these global challenges (Ologeanu-Taddei et al., 2024).

Based on the aforementioned, it could be stated that organizational sustainability encompasses the integration of economic, social, and environmental dimensions, often referred to as the "3Ps of business" – Profit, People and Planet – or the "triple bottom line" (Pislaru et al., 2019). The literature highlights a range of key elements associated with this concept, which are comprehensively summarized in Table 2 (Pislaru et al., 2019; Robu et al., 2023; Herghiligiu et al., 2021; Herghiligiu et al., 2024).

Starting from the aforementioned it could be mentioned that in the context of digital transformation and the global economy, corporate sustainability is evolving as a strategic approach that integrates advanced technologies to support environmental and social sustainability initiatives, while reducing the negative impact on natural resources and promoting circular business models (Ghobakhloo, 2020; Benitez et al., 2021). Thus, corporate sustainability becomes not only a compliance framework, but also a strategic differentiator in the contemporary competitive environment.

Table 2. Corporate sustainability – associated subdimensions (Pislaru et al., 2019; Herghiligiu et al., 2021; Robu et al., 2023; Herghiligiu et al., 2024).

No.	Dimension	Sub-dimension
1	Economic/ financial	▪ economic performance: profit, efficiency, and risk management ▪ workforce dynamics: employee motivation and ethics in management ▪ stakeholder engagement: suppliers' relationships and government relations ▪ strategic growth: innovation and marketing practices
2	Social	▪ workforce well-being: human resource programs, health and safety, and human rights ▪ community and social impact: local engagement, philanthropy, and volunteering ▪ customer-centric practices: consumer relationship management, product liability, and sustainable consumption ▪ operational excellence: quality management
3	Environmental	▪ resource efficiency: energy conservation, raw materials, and H_2O management ▪ environmental protection: waste management, pollution, and climate change mitigation ▪ sustainable operations: production management, supply chain management, and green suppliers ▪ environmental governance: environmental reporting, environmental management systems (EMS), and compliance with environmental requirements ▪ risk mitigation: environmental risk assessment and management

2.3 Digitalization role in organizational sustainability – literature limitation

Due to the complexity of organizational sustainability, the evaluation process has some limitations (Bolis et al., 2017). These limitations are associated with the classical and quantitative models used in the decision-making process (Hertin et al., 2009; Garrity, 2012; Maiorano and Savan, 2015), and the complexity of the sustainable development phenomenon (Foxon et al., 2013; Pislaru et al., 2019). The role of digitalization in sustainability represents a developing research area, full of opportunities, but also controversies. Literature in this field, even if it is at an early stage, recognizes the perspectives, still unexploited, that digital paradigms could bring to the development of organizational sustainability (Del Río Castro et al., 2021).

Likewise, literature presents some limitation regarding the role of digitalization in organizational sustainability, such as: lack of integrated frameworks that comprehensively address how digital technologies can simultaneously enhance economic, social, and environmental dimensions of sustainability (Ghobakhloo, 2020; Benitez et al., 2021); insufficient empirical evidence linking digital transformation to concrete sustainability outcomes (Ologeanu-Taddei et al., 2024); limited exploration of the paradoxes and tensions arising from the coexistence of sustainability objectives with economic and operational goals within digital ecosystems (Hanelt et al., 2021); inadequate understanding of the long-term impacts of digital technologies on environmental and social sustainability metrics (Favoretto et al., 2022; Benitez et al., 2021).

These literature gaps highlight the need for interdisciplinary research combining technological, managerial and sustainability perspectives to guide effective digital transformation strategies.

3. Research design

The research employs a statistical methodological approach (Jaba, 2002), which involves: (i) defining the target population and selecting the analyzed sample, (ii) identifying the relevant variables and determining the appropriate assessment model, and (iii) collecting and processing the data through statistical techniques in order to test the proposed model (Herghiligiu et al., 2019; Herghiligiu et al., 2024).

3.1 Analyzed population and sample

The studied population consists of national and international (private) organizations, while the proposed sample for analysis will include large organizations that (i) publicly disclose various specific data or sustainability reports and (ii) are publicly traded. Hence, in the final sample, based on relevant selection criteria (Herghiligiu et al., 2024), 6 very large industrial companies were selected predominantly from Northeastern Romania.

3.2 Proposed assessment model and associated variables

To fill this aforementioned literature gaps, the purpose of this research is to propose and to test a model that could asses the digitalization role in organizational sustainability – Figure 1 (Herghiligiu and Aevoae, 2024). In order to evaluate the link between these two main macro-dimensions ("digitalization" and "organizational

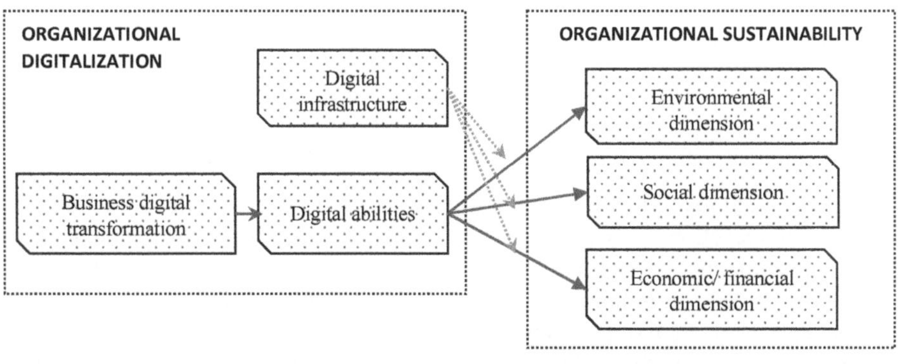

Figure 1. Proposed assessment model for evaluating the role of digitalization in organizational sustainability (Herghiligiu and Aevoae, 2024).

sustainability"), in the study we considered some relevant variables indicators that were/are reported or collected for a representative international automotive company.

Organizational digital transformation – associated variables/indicators (Figure 1), are described as followed (Gökalp and Martinez, 2021; Martínez-Peláez et al., 2023; Unlu et al., 2023; EC – DESI, 2023; Ologeanu-Taddei et al., 2024; Herghiligiu and Aevoae, 2024):

- *Business digital transformation*: Social media (Soc.Media.) – existence of profiles, accounts, and licenses associated with social media/number associated with blogs, sharing websites; Big Data (Big.D.) – percentage of big data usage/analysis at the organizational level from various sources; Cloud (cloud) – number of contracts/services purchased by the organization related to cloud computing (database hosting, accounting applications, project management); Electronic Information Sharing (PEI) – existence (and usage at the organizational level) of ERP (Enterprise Resource Planning) software to share information across different functional departments; Artificial Intelligence (AI) – level of AI usage within the organization; Online Commerce – % of electronic invoices issued compared to the total number of invoices (ConlineF) and % of online sales compared to the total sales made (ConlineV)/value of revenue from online commerce/% of revenue associated with online commerce;
- *Digital abilities*: Internet use (UNet) – level of internet usage by organizational human resources per week (hours) for performing work tasks; People with IT degrees (PD-IT) – number of organizational employees with an IT degree; Digital skills (average; above average; content creation) (Cdigit.) – individual level associated with performing informational tasks at the workplace/use of dedicated software (informational efficacy); IT organizational specialists (S-IT) – number of IT specialists within the organization; Training on increasing digital skills (TCAD) – number of training sessions at the organizational level aimed at enhancing digital skills of human resources;
- *Digital infrastructure*: Fixed infrastructure usage (Inf.Digit.Fix.) – number/type of subscriptions/contracts at the organizational level for access to digital infrastructure; Mobile infrastructure usage (Inf.Digit.Mob.) – number of mobile phones with subscriptions per employee.

Corporate sustainability – associated variables/indicators (Figure 1), are described as follows (Pislaru et al., 2019; Herghiligiu et al., 2021; Robu et al., 2023; Herghiligiu et al., 2024; Herghiligiu and Aevoae, 2024):

- *Economic/financial*: Stock price/Earnings per share (PE) – it reflects the financial market's reaction, as measured by the stock exchange rate, to the announcement of the company's net earnings relative to the number of issued shares; Financial return (ROE) – it is linked to the efficiency of capital utilization in relation to the net result; Economical return (ROA) – assets use efficiency in exploitation activities refers to the ability of an organization to optimize the utilization of its assets to generate maximum benefits; Return on sales (ROS) – sales efficiency based on recorded net margin evaluates how effectively an organization converts its sales revenue into profit; Return on investment (ROI) – it reflects the efficiency of corporate investments by evaluating the future cash flows generated, adjusted using the weighted average cost of capital; Gross margin on sales (OM) – sales efficiency based on recorded gross margin evaluates how effectively a company generates profitability from its core sales activities before accounting for operating expenses, taxes, and interest; Sales Growth Index (GS) – it is calculated as the ratio of the change in sales between two consecutive periods to the sales value in the reference period, providing a measure of the growth or decline in sales performance over time;

- *Environmental*: Water consumption (WaC) – organization's impact on water resources (m^3); Toxic metals (TM) – toxic metals discharged, weighted by the associated toxicity coefficient; Solid suspensions (SS) – average daily discharge rate of suspended solids (t. per year); Greenhouse gases (GHG) – encompasses CO_2 equivalent emissions (metric t.); Volatile organic compounds (VOC) – level of emissions related to volatile organic compounds; Sulfur dioxide and nitrogen oxides (SO_2 and NOx) – levels of SO_2 and NO_x emissions; Waste (OIW) – mass of solid waste disposed (t./unit); Hazardous Material (HIW) – the quantity of hazardous, harmful, or toxic materials utilized in production; Eco-design (ED) – percentage of fabricated products incorporating environmentally sustainable design principles; Recycling rate (RR) – percentage of finite products or recycled/discarded packaging;
- *Social*: Work shift (WS) – average number of hours worked per week; Salary (S) – average annual cost per employee; Sex ratio (SxR) – percentage of female employees relative to male employees; Professional training/development (PD) – percentage of employees who have attended training courses; Safety training (ST) – percentage of employees participating in safety and emergency procedures training; Work accidents (WA) – ratio of days with reported body injuries to the total number of working days; Local suppliers (LS) – ratio of local suppliers to the total number of suppliers; Local employees (LoE) – ratio of local employees to the total number of employees; Sustainable suppliers (SuS) – percentage of suppliers evaluated and selected based on their sustainability orientation.

4. Results and discussions

The following section presents the research results that aimed to test the proposed model in Figure 1. Hence a Structural Equation Modeling (SEM) analysis was performed, in order to evaluate the interconnections between digitalization and organizational sustainability. These findings also provide an understanding regarding relationships between proposed latent constructs and observed indicators. Likewise, it supports previous literature research that emphasize the transformative role of digital technologies in fostering sustainable organizational practices (Ghobakhloo, 2020; Favoretto et al., 2022).

The SEM analysis (performed with SmartPLS software) for assessing the digitalization role in organizational sustainability is presented in Figure 2 and corresponds to the assessment model proposed in Figure 1.

Studies such as Favoretto et al. (2022) indicate that digital abilities are essential for organizations to adapt to the digital changes. Also, Ologeanu-Taddei et al. underline that digital abilities support the achievement of sustainability goals. Likewise, the SEM analysis (Figure 2) identifies that with a high coefficient (0.984), *digital abilities* is the main catalyst of *business digital transformation* (0.990). Hence organizations need to use their resources on developing digital skills, including digitalization processes, technology integration and technical adaptability.

Digital infrastructure is essential for scaling technological innovations and integrating sustainability into business strategies (Brenner, 2018). Using digital infrastructure for big data analysis (Benitez et al., 2021) increases the efficiency of strategic decisions and contributes to economic sustainability. However, the coefficients of 0.182 and 0.115 (Figure 2) indicate a moderate influence of *digital infrastructure.*

Economic performance (ROA, ROE) is a key element in digitalization investments (Hanelt et al., 2021). Ologeanu-Taddei et al. (2024) show that economic profitability facilitates the implementation of digital and environmental initiatives. The coefficient of 0.429 (Figure 2) between the *economic dimension* and *digital abilities* indicates a representative link. Organizations should improve their financial performance by supporting projects associated with digitalization.

Ghobakhloo (2020) indicates that digitalization can reduce environmental impact through resource and emissions management. Brenner (2018) emphasizes that digital technologies enable the development of circular economies. Improving *environmental management* (environmental impact) (coefficient 0.384 –

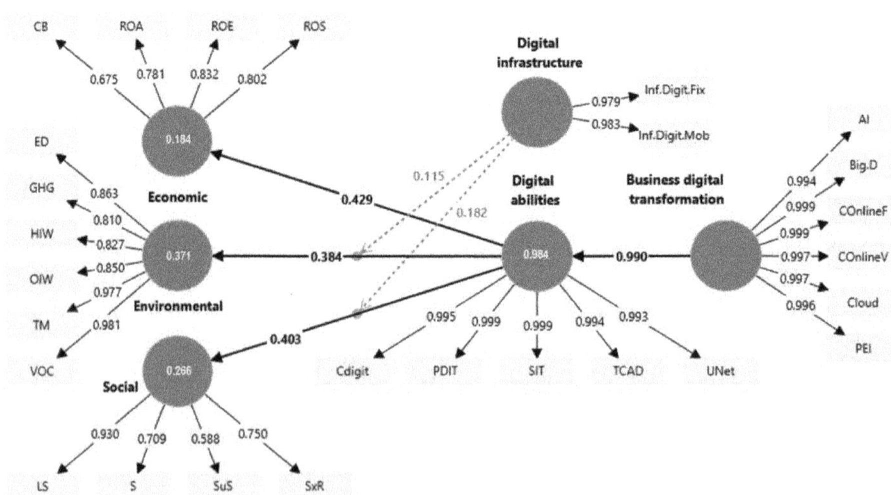

Figure 2. Assessment model for evaluating the role of digitalization in organizational sustainability – SEM analysis (Herghiligiu and Aevoae, 2024).

Figure 2) can be a significant outcome of *digital abilities*. Organizations can implement solutions such as IoT to monitor and optimize energy consumption and reduce toxic waste.

Corporate *social responsibility* plays a significant role in supporting communities and attracting local partners (Kantabutra and Ketprapakorn, 2020). The model coefficient of 0.403 (Figure 2) essentially indicates a strong relationship between *social* impact and *digitalization*. Consequently, organizations would be well advised to support local suppliers and implement social sustainability initiatives, integrating digital solutions for transparency and inclusion.

Hence the results align with the theoretical framework established by Vial (2019) and Hanelt et al. (2021), emphasizing the multidimensional impact of digital transformation on economic, social, and environmental outcomes. Special attention is given to the central role of digital abilities as a driver of business transformation (Ologeanu-Taddei et al., 2024; Benitez et al., 2021). Also, the results are contextualized within the broader implications for organizational practices. The findings resonate with Brenner (2018), who argued that digital infrastructure and abilities are key enablers for integrating sustainability into business strategies. This section bridges theoretical constructs with practical applications, providing actionable insights for decision-makers seeking to align digital transformation efforts with sustainability objectives.

5. Conclusions

Research results highlight that digital skills have the greatest impact on digital transformation, and the literature supports that they are fundamental for organizational success in the digital age. Developing organizational digitalization enhances economic, environmental and social objectives and supports sustainability performance monitorization. Initiatives associated with digitalization can optimize the resource use and reduce costs through process automation and Enterprise resource planning integration. The implementation of digital platforms can easily allow the collection, analysis and reporting of sustainability data, an extremely important aspect for actual competitive environment.

The main research results, obtained from SEM analysis – model (Figure 2) developed based on literature, confirm and underline that digitalization serves as a critical enabler of organizational sustainability by fostering economic efficiency, reducing environmental impact and enhancing social responsibility. By leveraging digital abilities, robust infrastructures and advanced technologies, organizations can integrate sustainability into their core strategies, achieving resilience and alignment with global sustainability requirements.

References

Annunziata, M. and Biller S. (2015). The Industrial Internet and the future of work, ASME. Mechanical Engineering 137(09). https://doi.org/10.1115/1.2015-Sep-1.

Baiyere, A., Berente, N. and Avital, M. (2023). On digital theorizing, clickbait research, and the cumulative tradition. Journal of Information Technology 38. https://doi.org/10.1177/02683962231153940.

Bansal, P. and Roth, K. (2000). Why companies go green: A model of ecological responsiveness. Academy of Management Journal 43(4). https://doi.org/10.2307/1556363.

Benitez, E. O., da Costa, M. B., Baierle, I. C., Schaefer, J. L., Benitez, G. B., Lima do Santos, L. M. A. et al. (2021). Expected impact of Industry 4.0 technologies on sustainable development: A study in the context of Brazil's plastic industry. Sustainable Production and Consumption 25. https://doi.org/10.1016/j.spc.2020.07.018.

Birkel, H. and Müller, J. M. (2021). Potentials of industry 4.0 for supply chain management within the triple bottom line of sustainability – a systematic literature review. Journal of Cleaner Production 289. https://doi.org/10.1016/j.jclepro.2020.125612.

Bolis, I., Morioka, S. N. and Sznelwar, L. I. (2017). Are we making decisions in a sustainable way? A comprehensive literature review about rationalities for sustainable development. Journal of Cleaner Production 145. https://doi.org/10.1016/j.jclepro.2017.01.025.

Brenner, B. (2018). Transformative sustainable business models in the light of the digital imperative—a global business economics perspective. Sustainability 10(12). https://doi.org/10.3390/su10124428.

Brenner, B. and Hartl, B. (2021). The perceived relationship between digitalization and ecological, economic, and social sustainability. Journal of Cleaner Production 315. https://doi.org/10.1016/j.jclepro.2021.128128.

Del Río Castro, G., González Fernández, M. C. and Uruburu Colsa, A. (2021). Unleashing the convergence amid digitalization and sustainability towards pursuing the Sustainable Development Goals (SDGs): A holistic review. Journal of Cleaner Production 280(1). https://doi.org/10.1016/j.jclepro.2020.122204.

Elkington, J. (1994). Towards the sustainable corporation: Win-win-win business strategies for sustainable development. California Management Review 36(2). https://doi.org/10.2307/41165746.

European Commission (EC – DESI). (2023). Digital Economy and Society Index (DESI) 2023. Publications Office of the European Union. https://digital-strategy.ec.europa.eu/en/policies/desi.

European Commission (EC), European Skills Agenda for Sustainable Competitiveness, Social Fairness and Resilience. (2020). Retrieved from https://ec.europa.eu/commission/presscorner/detail/en/ip_20_1196.

Favoretto, C., Mendes, G. H. d. S., Filho, M. G., Gouvea de Oliveira, M. and Ganga, G. M. D. (2022). Digital transformation of business model in manufacturing companies: Challenges and research agenda. Journal of Business & Industrial Marketing 37(4). https://doi.org/10.1108/JBIM-10-2020-0477.

Foxon, T. J., Kohler, J., Michie, J. and Oughton, C. (2013). Towards a new complexity economics for sustainability. Cambridge Journal of Economics 37. https://www.jstor.org/stable/23601929.

Garrity, E. J. (2012). Tragedy of the commons, business growth and the fundamental sustainability problem. Sustainability 4. https://doi.org/10.3390/su4102443.

Gebhardt, C. (2017). Humans in the loop: The clash of concepts in digital sustainability in smart cities. pp. 85–93. *In*: Osburg, T. and Lohrmann, C. (eds.). Sustainability in a Digital World: New Opportunities Through New Technologies. Springer International Publishing. https://doi.org/10.1007/978-3-319-54603-2_7.

George, G., Merrill, R. K. and Schillebeeckx, S. J. D. (2021). Digital sustainability and entrepreneurship: How digital innovations are helping tackle climate change and sustainable development. Entrepreneurship Theory and Practice 45(5). https://doi.org/10.1177/1042258719899425.

Ghobakhloo, M. (2020). Industry 4.0, digitization, and opportunities for sustainability. Journal of Cleaner Production 252. https://doi.org/10.1016/j.jclepro.2019.119869.

Gilchrist, A. (2016). Industry 4.0: the Industrial Internet of Things. Apress, Springer.

Gökalp, E. and Martinez, V. (2021). Digital transformation capability maturity model enabling the assessment of industrial manufacturers. Computers in Industry 132: 103522. https://doi.org/10.1016/j.compind.2021.103522.

Gökalp, M. O., Gökalp, E., Kayabay, K., Koçyiğit, A. and Eren, P. E. (2021a). Data-driven manufacturing: An assessment model for data science maturity. Journal of Manufacturing Systems 60. https://doi.org/10.1016/j.jmsy.2021.07.011.

Gökalp, M. O., Kayabay, K., Gökalp, E., Koçyiğit, A. and Eren, P. E. (2021b). Assessment of process capabilities in transition to a data-driven organisation: A multidisciplinary approach. IET Software 15. https://doi.org/10.1049/sfw2.12033.

Gouvea, R., Kapelianis, D. and Kassicieh, S. (2018). Assessing the nexus of sustainability and information & communications technology. Technological Forecasting and Social Change 130. https://doi.org/10.1016/j.techfore.2017.07.023.

Guandalini, I. (2022). Sustainability through digital transformation: A systematic literature review for research guidance. Journal of Business Research 148. https://doi.org/10.1016/j.jbusres.2022.05.003.

Hahn, T., Figge, F., Pinkse, J. and Preuss, L. (2018). A paradox perspective on corporate sustainability: descriptive, instrumental, and normative aspects. Journal of Business Ethics 148. https://doi.org/10.1007/s10551-017-3587-2.

Hanelt, A., Bohnsack, R., Marz, D. and Antunes Marante, C. (2021). A systematic review of the literature on digital transformation: insights and implications for strategy and organizational change. Journal of Management Studies 58. https://doi.org/10.1111/joms.12639.

Herghiligiu, I. V., Robu, I. B., Pislaru, M., Vilcu, A., Asandului, A. L., Avasilcai, S. et al. (2019). Sustainable EMS integration and business performance: A balance assessment approach using fuzzy logic. Sustainability 11(19): 5311. https://doi.org/10.3390/su11195311.

Herghiligiu, I. V., Robu, I. B., Vilcu, A. and Pislaru, M. (2021). Organizational sustainability main components identification using PCA. Proceedings of the 7th International Symposium TTPF 2021, Iasi, Romania. https://doi.org/10.2478/9788366675735-041.

Herghiligiu, I. V. and Aevoae, G. M. (2024). Evaluation of the Role of Digitalization in the Context of Organizational Sustainability Based on an Innovative Framework of Statistical Methods and Soft Computing Techniques, Research Project Report no. 3, Research Projects of the Romanian Academy of Scientists (AOSR) for young researchers, "AOSR-TEAMS-II" 2023–2024 Edition – "Digital Transformation in Sciences". https://www.aosr.ro/wp-content/uploads/2024/08/10-Rap-Cerc-2023-2024-etapa-3.pdf.

Herghiligiu, I. V., Robu, I. B., Vilcu, A., Pislaru, M., Pohonu-Dragomir, S. C., Cojocaru, M. et al. (2024). Organizational sustainability score – probability approach using fuzzy logic. Environmental Engineering and Management Journal 23(2). https://doi.org/10.30638/eemj.2024.019.

Hertin, J., Jacob, K., Pesch, U. and Pacchi, C. (2009). The production and use of knowledge in regulatory impact assessment—an empirical analysis. Forest Policy and Economics 11. https://doi.org/10.1016/j.forpol.2009.01.004.

Jaba, E. (2002). Statistica, 3rd Edition, Economica Publishing House, Bucharest, Romania.

Kantabutra, S. and Ketprapakorn, N. (2020). Toward a theory of corporate sustainability: A theoretical integration and exploration. Journal of Cleaner Production 270(10). https://doi.org/10.1016/j.jclepro.2020.122292.

Kiron, D. and Unruh, G. (2018). The convergence of digitalization and sustainability. Big Idea: Leading Sustainable Organizations, MIT Sloan Management Review. https://sloanreview.mit.edu/article/the-convergence-of-digitalization-and-sustainability/.

Kohler, J., Geels, F., Kern, F., Markard, J., Wieczorek, A., Alkemade, F. et al. (2019). An agenda for sustainability transitions research: state of the art and future directions. Environmental Innovation and Societal Transitions 31. https://doi.org/10.1016/j.eist.2019.01.004.

Maiorano, J. and Savan, B. (2015). Barriers to energy efficiency and the uptake of green revolving funds in Canadian universities. International Journal of Sustainability in Higher Education 16(2). https://doi.org/10.1108/IJSHE-07-2012-0062.

Markus, M. L. and Rowe, F. (2023). The digital transformation conundrum: labels, definitions, phenomena, and theories. Journal of the Association for Information Systems 24. https://doi.org/10.17705/1jais.00809.

Martínez-Peláez, R., Ochoa-Brust, A., Rivera, S., Félix, V. G., Ostos, R., Brito, H. et al. (2023). Role of digital transformation for achieving sustainability: mediated role of stakeholders. Key Capabilities, and Technology, Sustainability 15(14). https://doi.org/10.3390/su151411221.

Nguyen, N. P. and Hoai, T. T. (2022). The impacts of digital transformation on data-based ethical decision-making and environmental performance in Vietnamese manufacturing firms: The moderating role of organizational mindfulness. Cogent Business & Management 9(1): 1–20. https://doi.org/10.1080/23311975.2022.2101315.

Ologeanu-Taddei, R., Hönigsberg, S., Weritz, P., Wache, H., Mittermeier, F., Tana, S. et al. (2024). The relationship of digital transformation and corporate sustainability: Synergies and tensions. Technological Forecasting & Social Change 210(2025). https://doi.org/10.1016/j.techfore.2024.123809.

Osburg, T. and Lohrmann, C. (eds.). (2017). Sustainability in a Digital World. New Opportunities through New Technologies. Springer. https://doi.org/10.1007/978-3-319-54603-2.

Pislaru, M., Herghiligiu, I. V. and Robu, I. B. (2019). Corporate sustainable performance assessment based on fuzzy logic. Journal of Cleaner Production 223. https://doi.org/10.1016/j.jclepro.2019.03.130.

Robu, I. B., Aevoae, G. M., Mardiros, D. N. and Herghiligiu, I. V. (2023). Mergers & acquisition decisions in the energy sector based on financial transparency and audit opinions. Eastern European Economics 1–3. https://doi.org/10.1080/00128775.2023.2225484.

Sachs, J. D., Schmidt-Traub, G., Mazzucato, M., Messner, D., Nakicenovic, N. and Rockström, J. (2019). Six transformations to achieve the sustainable development goals. Nature Sustainability 2. https://doi.org/10.1038/s41893-019-0352-9.

Schuh, G., Anderl, R., Gausemeier, J., ten Hompel, M. and Wahlster, W. (2017), Industrie 4.0 Maturity Index Managing the Digital Transformation of Companies, Munich Herbert Utz.

Seele, P. and Lock, I. (2017). Game-changing potential of digitalization for sustainability: possibilities, perils, and pathways. Sustainability Science 12. https://doi.org/10.1007/s11625-017-0426-4.

Seidel, S., Bharati, P., Fridgen, G., Watson, R. T., Albizri, A., Boudreau, M.-C. et al. (2017). The sustainability imperative in information systems research. Communications of the Association for Information Systems 40. https://doi.org/10.17705/1CAIS.04003.

Swan, M. (2015), Blockchain: Blueprint for a New Economy, O'Reilly Media, Inc.

United Nations (UN), Department of Economic and Social Affairs, Sustainable Development, The 17 Goals. (2023a). https://sdgs.un.org/goals.

United Nations (UN), UN Environment Programme: UNEP. (2023b). https://www.unep.org/.

Unlu, H., Demirörs, O. and Garousi, V. (2023). Readiness and maturity models for Industry 4.0: A systematic literature review. Journal of Software-Evolution and Process. https://doi.org/10.1002/smr.2641.

Verhoef, P. C., Broekhuizen, T., Bart, Y., Bhattacharya, A., Dong, J. Q., Fabian, N. et al. (2021). Digital transformation: A multidisciplinary reflection and research agenda. Journal of Business Research 122: 889–901. https://doi.org/10.1016/j.jbusres.2019.09.022.

Vial, G. (2019). Understanding digital transformation: A review and a research agenda. Technological Forecasting & Social Change 146. https://doi.org/10.1016/j.jsis.2019.01.003.

Walker, J., Pekmezovic, A. and Walker, G. (2019). Sustainable Development Goals: Harnessing Business to Achieve the SDGs through Finance, Technology and Law Reform. Wiley.

Wessel, L., Baiyere, A., Ologeanu-Taddei, R., Cha, J. and Blegind Jensen, T. (2021). Unpacking the difference between digital transformation and IT-enabled organizational transformation. Journal of the Association for Information Systems 22. https://doi.org/10.17705/1jais.00655.

Zheng, Y. and Walsham, G. (2021). Inequality of what? An intersectional approach to digital inequality under Covid-19. Information and Organization 31(1). https://doi.org/10.1016/j.infoandorg.2021.100341.

Chapter 3

Strategies for Improving Organizational Capacity for the Circular Economy

Timea Agache,[1,*] *Muddassar Sarfraz,*[2] *Andrei Agache*[1] and *Alexandra Coroian*[1]

1. Introduction

The shift to a Circular economy is no longer a desire but a necessity due to global issues such as resource depletion, environmental degradation, and unsustainable production and consumption. Companies have to prevent these issues by developing strategies that reduce waste, enhance resource efficiency, and establish closed-loop systems. However, a precondition for this is organizational capacity, which denotes an entity's capacity to modify internal processes, promote innovation, involve stakeholders, and cultivate the requisite skills, resources, and technology. On the contrary, insufficient organizational capability may hinder businesses from integrating circular processes, thus jeopardizing their economic and environmental sustainability objectives. Therefore, understanding strategies to enhance organizational capacity is crucial for firms to transition to the Circular economy.

This paper aims to identify the relationship between Circular economy and organizational capabilities. The focus is to identify: (1) main obstacles of the fundamental categories related to organizational capacity, and (2) strategies that can enhance the identified capacities. The report sought to establish a conceptual

[1] Politehnica University of Timisoara, Faculty of Management in Production and Transportation, 14 Remus Street, Timisoara, Romania.
[2] School of Management, Zhejiang Shuren University, Hangzhou 310015, Zhejiang, PR China.
Email: muddassar.sarfraz@gmail.com
* Corresponding author: timea.cisma@student.upt.ro

DOI: 10.1201/9781003518990-3

framework to assist businesses in enhancing their ability to effectively employ circular practices. This framework will function as an essential instrument for forthcoming organizational development and strategic planning, hence facilitating overarching environmental and economic sustainability objectives.

2. Literature review

Circular Economy is an innovative approach to achieve sustainable development; yet, some research supports earlier authors' claims that its relationship with sustainable development is weak. While many folks view the Circular economy as a pathway to economic prosperity, prior scholars doing narrative literature reviews have emphasized its environmental goals. This gap is particularly evident in some definitions, where economic prosperity is of utmost importance and social issues are often neglected, highlighting that this dimension remains insufficiently analyzed (Kirchherr and Piscicelli, 2019).

The Circular economy is frequently viewed as a promising concept, especially for its capacity to engage the business sector in sustainable development initiatives. Many academics approve that resources extracted from nature and transformed into economically valued products or services have to be utilized repeatedly rather than only once. This linear economic model, defined by a "take-make-dispose" approach, results in significant waste and the exhaustion of natural resources. Contrary to this, circular economy promotes the ongoing use of resources, while reducing the need for raw materials, waste and emissions associated with economic activity (Korhonen et al., 2018).

The circular principles offer a sustainable framework for tackling global concerns. Surely, these principles are mainly known for reevaluating conventional methods of production and consumption. The implementation of these mitigates greenhouse gas emissions and saves resources, which leads to better air and water quality. Moreover, the circular economy stimulates innovation, diminishes resource reliance, and enhances resilience through collaboration among governments, industries, and academics, providing a route to a more sustainable and efficient future (Yang et al., 2023).

While developing circular strategies, organizations must take into account multiple interconnected aspects that could affect a business. Initially, products and services have to be highlighted that extend product lifecycles, incorporate product-service systems, and enable collaborative or virtual consumption. This also means motivating consumers to return used products to enable resource recirculation. Another critical aspect is that all strategies have to be aligned with the values of the customer segments, in order to ensure a strong correspondence between the offerings and the needs of the clients. Moreover, the correct choice of channel is crucial for delivering value, including digital methods for marketing and distribution, as well as enabling consumer engagement. It has been discovered that consumer interactions must emphasize flexibility, incorporate manufacturing that aligns with consumer demand, and foster community partnerships. Regarding the revenue streams, it should derive from the value delivered by the offerings, either through performance-based models such as pay-per-use or via direct payments for products or services. Moreover, revenues may also include the value derived from reintegrating goods into the supply chain. Moreover, the cost structure has to cover the financial implications of these circular processes, including consumer incentives and the requirement for specialized accounting techniques to evaluate costs appropriately. Last but not the least, an efficient take-back system incorporating channels and consumer engagement is essential to facilitate product returns. In terms of operational activities, it should focus on boosting efficiency through improved procedures, technology developments, and innovative product design (Lewandowski, 2016).

As mentioned before, organizational capacity is crucial for the implementation of circular economy concepts, since it integrates internal competencies with external elements. This state of balance allows organizations to adopt circular business models, providing a strategic, long-term framework for resource optimization, stakeholder engagement, and sustainable value creation. The application of circular strategies occurs in three primary areas: production, product innovation, and recycling practices, each influenced by internal and external factors. Indeed, organizational capacity enables organizations to strategically match internal resources with external demands, hence promoting the incorporation of circular practices into their operations (Castro-Lopez et al., 2023).

3. Methodology

The primary aim of this study was to examine the relationship between Circular Economy practices and organizational capabilities. The study was structured around two principal objectives:

- *Main obstacles of the fundamental categories related to organizational capacity*: The first step was to identify which organizational capacities are directly linked to implementation of circular practices and which are the main obstacles of each cluster. In order to found out, prior studies, articles, and frameworks on this topic were analyzed.
- *Strategies that can enhance the identified capacities*: In the second phase, our research focused on identifying methods which enhance the previously established organizational capabilities in the context of circular practices. This was achieved by linking strategies commonly used in circular practices with the organizational capabilities they enhance.

4. Capacities of circular economy implementation

The barriers to implement the Circular economy (Araujo Galvão et al., 2018; Govindan and Hasanagic, 2018; Kirchherr et al., 2018) can be categorized into five clusters. These clusters are essential for the successful implementation of circular economy principles. First, only business with financial capacity as in terms of resources and investment can be successful. Second, technological and operational capacities are essential for the adoption of innovative technologies and processes that facilitate circular strategies such as recycling, reuse, or resource recovery. Third, governance and culture are crucial, as robust leadership and dedication to circular concepts can facilitate the necessary cultural transformation. Fourth, collaboration with other stakeholders are essential for addressing resource or skill deficiencies in the value chain. Last but not least, knowledge and process management are crucial for efficiently handling data and procedures associated with product life cycles and resource management. Should these obstacles be addressed, a company can exhibit the ability to execute and maintain circular economy principles.

4.1 Financial capacity

The financial challenges linked to implementing circular practices are significant, particularly for businesses transitioning from a traditional linear approach. A major obstacle is the large upfront investment required to implement circular models, which presents a huge strain for micro and small enterprises due to their limited financial resources. Furthermore, the complexity of reorganizing business activities to enable circular models raises financial burden. These characteristics expose companies to financial risks, as the circular economy is still developing. A major challenge is the absence of financial, organizational, and national benchmarks to evaluate the progress of circular firms. Before achieving financial success, firms must assess costs at each phase of manufacturing, taking into account the entire product longevity. Therefore, efficient accounting systems are crucial for tracking the costs, as different resources may have distinct life cycles. In this context, financial incentives, subsidies, and increased knowledge among governments, businesses, and consumers are essential for development and success of the circular economy (Gonçalves et al., 2022).

The adoption of circular practices necessitates substantial financial expenditure, with the degree of investment closely related to the level of its implementation in organizations. The accessibility and quality of financial resources are important elements in ascertaining the effectiveness of projects. Without a doubt, organizations with access to more capital are better equipped to execute various circular initiatives. Additionally, the quality of these resources is also significant because financing circular operations typically necessitates greater collateral compared to other types of investment. Moreover, public grants and subsidies are catalysts for advancing the circular economy, because they mitigate the financial obstacles encountered by the companies. The presence of public financial assistance and environmental research and development affect the extent to which companies engage in the circular economy (Aranda-Usón et al., 2019).

The national financial environment surely impacts the accessibility of money and investment for the circular business models. An unfavorable one is characterized by inadequate economic performance, high taxation, or fluctuating exchange rates. All these might hinder the flow of capital into circular economy projects. On the contrary, an enabling environment, which is supported by favorable tax policies, government funding, and regulatory frameworks, enhances the ability of an organization to attract investments. Thus, firms proficient in assessing liquidity, improving asset quality, and integrating sustainable practices into their financial plans, are more likely to attract investments in the context of circular economy (Agyapong and Tweneboah, 2023).

4.2 Operational and technological capacity

The operational capacity of a circular economy has the duty to harmonize sustainability while guaranteeing long-term profitability. Therefore, sustainability involves the examination of the production-consumption system as a cohesive entity within the context of circular economy. However, circularity underscores the importance of maintaining resources in utilization via closed-loop production and consumption methodologies. Resilience is the foundation of operational capacity. It refers to the system's ability to evolve, integrate elements that enhance and expedite processes while conforming to environmental sustainability principles (Suárez-Eiroa et al., 2021).

There are a series of techniques that can ensure operational capacity. First, effective resource use and waste reduction can be ensured by maximizing the retention of resources within the system. Second, creating products and systems with a focus on circularity enhances this goal. Third, concluding the system is a crucial concept that underscores the importance of reusing and recycling materials to reduce reliance on external resources. Moreover, synchronizing regeneration rates with environmental absorption capacities guarantees ecological equilibrium. Similarly, minimizing the system's total size decreases resource requirements and mitigates environmental effect (Suárez-Eiroa et al., 2019).

Last but not least, technology increases operational capacity. The primary technological capacity relevant to the circular economy are blockchain technology and Industry 4.0. Blockchain technology facilitates the advancement of circular economy activities and improves essential functions such as circular procurement, circular design, recycling, and remanufacturing while offering transparency, traceability, and efficiency throughout supply chains. Moreover, blockchain technology guarantees material authenticity, promotes stakeholder trust, and aids compliance with circular standards through secure data sharing and real-time tracking (Wynn and Jones, 2022). Industry 4.0 technology are essential for advancing circular practices because it enhance activities, which are consequently associated with improved environmental and operational results. The implementation of circular processes has a beneficial impact on a firm's environmental performance, economic efficiency, and operational performance (Khan et al., 2022).

4.3 Governance and cultural values

Government support and organizational culture can significantly influence the enhancement of sustainable performance. For instance, government fundings indirectly affect sustainable performance by fostering eco-innovation and circular behaviors, including resource efficiency, waste reduction, and recycling. Similarly, corporate culture influences sustainable performance both directly and indirectly, via eco-innovation and circular economy activities. Therefore, it can be said that institutional variables, such as governmental regulations and organizational culture, are key capacities of sustainability success (Ali et al., 2024).

Proactive organizations in the context of circular economy often exhibit superior management techniques, driven by the dedication to foster a sustainable corporate culture. The engagement of a firm's upper management is essential in this change, since top executives with sustainability education and a deep understanding of circular economy principles can significantly influence performance. In particular, organizations aiming for advancement in this domain should incorporate sustainability into their culture, guided by senior management exemplifying circular ideas and practices. Due to the fact that organizations may exist at various maturity levels in their circular economy implementation, managers and policymakers have to assess the current position and formulate strategies for proactive improvement (Sehnem et al., 2019).

4.4 Partnership and engagement

Organizational capacity can be analyzed via the perspective of individual-level micro foundations and their connection to social transformation and emotional organizational commitment. Individual-level micro foundations—such as personal values, abilities, and behaviors—are positively correlated with the overarching concept of social change and the emotional attachment employees experience towards their organization, referred to as affective organizational commitment. Social transformation has a direct positive effect on emotional organizational commitment, creating a workplace climate that connects individual purpose with organizational objectives. Moreover, social change partially mediates the connection between micro foundations and organizational commitment, suggesting that cultivating an environment conducive to social change can strengthen the link between individual contributions and total commitment (Ghouri et al., 2019).

Increased diversity, concerning the linked social sectors, and prolonged partnerships enhance sustainability initiatives; nonetheless, this impact is exclusively seen in responsive firms—those actively mitigating environmental pollutants. Conversely, non-responsive firms exhibited diminished progress in sustainability despite heightened diversity and extended relationship durations. This indicates that the nature and extent of partnerships, especially regarding diversity, significantly impact sustainability outcomes, with companies that are more attuned to environmental concerns achieving superior results from diversified and enduring collaborations (Valbuena-Hernandez and Ortiz-de-Mandojana, 2022).

Circular economy approaches significantly enhance organizational resilience by promoting resilience capacities, adaptability, and collaboration. These elements enhance the internal framework of businesses and positively affect network resilience and robustness, especially within the realm of industrial symbiosis. Organizations that more extensively employ circular economy principles demonstrate enhanced long-term resilience relative to their counterparts. This indicates that Circular economy practices not only bolster an individual organization's capacity to endure disturbances but also augment the resilience of the larger network in which they function (Karman et al., 2024).

4.5 Human resources and workforce

The development of human resource development is crucial for the success of businesses of all sizes, as it is widely recognized that no enterprise can prosper without human capital. A key function of it is to engage personnel in a manner that stimulates sales growth and profitability. Nonetheless, obstacles like insufficient funds might hinder the efficacy of training methodologies, as evidenced in certain organizations that cannot execute role-playing exercises due to fiscal limitations. Employing competent personnel markedly improves productivity, highlighting the significance of professional human resources in optimizing the use of other organizational assets. Furthermore, effective human resources methods, such as training and motivational aids, substantially enhance employee performance in all sectors (Okoye and Ezejiofor, 2013).

The ability for human resource and skill development is essential for fostering organizational innovation and improving knowledge management. Strategic human resource practices, including compensation and training, substantially enhance organizational creativity. Compensation and training are critical determinants of an organization's capacity to innovate. The capacity for knowledge management is affected by performance appraisal and remuneration systems, which are significant determinants of a company's ability to manage and utilize knowledge efficiently. These observations underscore the necessity of connecting human resource strategies with business objectives to foster innovation and improve knowledge management capabilities (Afacan Fındıklı et al., 2015).

Moreover, human resource policies and practices within a strategic human resource management framework substantially affect individual attitudes and behaviors, which collectively enhance organizational resilience. Recruitment, training, and performance management influence employee engagement with organizational objectives, promoting adaptability and collaboration (Lengnick-Hall et al., 2011).

5. Strategies to develop organizational capacity

Building on the list of strategies identified by Salvador et al., we aimed to explore the relationship between these strategies and organizational capacity. These strategies can significantly improve an organization's ability to adapt, innovate, and perform efficiently by strengthening key organizational capacities such as human resources, processes, infrastructure, and knowledge. The strategies include: "establishing strategic partnerships for circularity and engaging stakeholders throughout the value chain, designing for circularity, eliminating waste through design, fostering industrial symbiosis, promoting reuse, recycling, and reconditioning, adopting environmentally driven material usage practices, extending product lifespan, implementing take-back systems (TBS), integrating product-service systems (PSS), enabling refurbishment, remanufacturing, repair and maintenance, embracing dematerialization, and utilizing digital technologies to support circularity" (Salvador et al., 2021). Furthermore, some academics have differentiated between the strategy of Designing long-life products and Design for product-life extension. The former pertains to guaranteeing that items have superior quality and durability, whilst the latter emphasizes the potential for prolonging the product's lifespan (Bocken et al., 2016).

The table (Table 1 - Connection between circular strategies and organizational capacities) elucidates the correlation between diverse circular strategies and organizational capacities, demonstrating the impact of each strategy on organizational capacities, including financial, operational and technological, governance and culture, partnership and engagement, and human resources and workforce.

Table 1. Connection between circular strategies and organizational capacities (Adapted from Salvador et al. (2021) with additional analysis and modifications by the author.)

Strategy	Financial Capacity	Operational and Technological Capacity	Governance and Culture	Partnership and Engagement	Human Resources and Workforce
Developing strategic partnerships for circularity		X	X	X	
Engaging stakeholders along the value chain	X	X	X	X	X
Design for circularity		X			
Designing out waste		X			
Industrial symbiosis		X	X	X	
Reuse		X			
Recycling		X			
Refurbishment		X			
Remanufacturing		X			
Repair and maintenance		X			
Environmentally friendly material		X			
Designing long-life products		X			
Design for product-life extension		X			
Take-back systems	X	X		X	
Product-service systems	X	X		X	X
Dematerialization		X			
Digital technologies to enable circularity	X	X			X
Educating for circularity	X		X	X	X

Numerous tactics predominantly influence operational and technological capabilities, including "Designing out waste", "Industrial symbiosis" and "Digital technologies to enable circularity". These techniques generally necessitate substantial investments in innovative technology, processes, and systems that maximize resource utilization, minimize waste, and improve efficiency. "Industrial symbiosis" promotes collaboration among enterprises to exchange resources or waste. Nevertheless, these actions need operational and technological coordination. On the other hand, "Digital technologies to enable circularity" necessitates that firms invest in technological infrastructure and tools to oversee and enhance circular processes. Consequently, improved operational and technological capacities allow firms to enhance resource efficiency and sustainability.

Although operational and technological capacities are vital for the development of a circular economy, it is necessary to recognize that other competencies are just as essential. Governance and culture can be established by "Establishing strategic partnerships for circularity" and "Educating for circularity". These strategies require collaboration with external stakeholders and enable circular thinking. In addition, "Engaging stakeholders throughout the value chain" creates a collaborative governance framework, transforming the corporate culture towards transparency and enduring relationships. Moreover, training on circular practices is essential for cultivating a workforce that is informed and connected with the principles of the circular economy, thereby promoting a sustainable culture.

Financial capacity is affected by "Engaging stakeholders along the value chain" and "Product-service systems" as these approaches create long-term financial sustainability. In addition, "Take-back systems" necessitate that enterprises allocate substantial financial resources towards reverse logistics systems. Nevertheless, although these strategies emphasize financial elements, their immediate impact on financial capacity is less apparent than that of operational and technological strategies.

Partnership and engagement are significantly shaped by initiatives that encompass external collaboration, including "Developing strategic partnerships for circularity", "Industrial symbiosis" and "Take-back systems". These tactics necessitate robust networks and collaboration with diverse stakeholders, ranging from suppliers to consumers. "Industrial symbiosis" promotes resource sharing among companies, fostering reciprocal value and improving collaborative capacity. Product-service systems cultivate enduring ties between enterprises and consumers, which are vital for establishing sustainable partnerships.

Ultimately, human resources and the workforce are influenced by initiatives such as "Educating for circularity" and "Design for circularity". These initiatives emphasize the development of essential skills and knowledge within the workforce to properly apply circular processes. "Design for circularity" necessitates that employees comprehend sustainable product design ideas, while "Educating for circularity" guarantees that the entire firm adopts the practices of the circular economy. Strategies like "Repair and maintenance" and "Remanufacturing" rely on specific worker capabilities, necessitating ongoing training and development.

Meaning, all these strategies are influenced by human resources and workforce capacity.

6. Conclusions

The successful application of Circular economy is a transformative approach to address environmental and economic challenges. This research has highlighted the significance of organizational capability in the implementation of circular strategies. Understanding strategies to enhance organizational capacity is crucial for firms to transition to the Circular economy.

The main capacities that an organization needs to have in order to successfully implement circular practices are: financial capacity, technological and operational capacity, governance and culture, partnership and engagement, and human resource and workforce. Each of these capacities presents distinct obstacles related to: high-quality capital, blockchain technologies, Industry 4.0, leadership commitment, cultural alignment, innovation and collaboration.

The conceptual framework presents the relationship between circular strategies and which capacities could be improved through their implementation. Financial capacity can be enhanced by involving stakeholders along the value chain, implementing industrial symbiosis, establishing take-back systems, and adopting product-service systems. Operational and technological capacity is augmented by stakeholder engagement, circular design, waste elimination, industrial symbiosis, reuse, recycling, refurbishment, remanufacturing, repair and maintenance, eco-friendly materials, long-life product design, product-life extension strategies, take-back systems, product-service systems, and digital technologies. Governance and culture can be enhanced by establishing strategic partnerships for circularity, involving stakeholders, promoting industrial symbiosis, implementing take-back systems, adopting product-service systems, and providing education for circularity. Enhancing cooperation and involvement involves cultivating strategic alliances for circularity, including stakeholders, fostering industrial symbiosis, implementing take-back systems, establishing product-service systems, leveraging digital technologies, and promoting education for circularity. Human resources and workforce development are facilitated by involving stakeholders and promoting education for circularity.

References

Afacan Fındıklı, M., Yozgat, U. and Rofcanin, Y. (2015). Examining organizational innovation and knowledge management capacity the central role of strategic human resources practices (SHRPs). Procedia - Social and Behavioral Sciences 181: 377–387. https://doi.org/10.1016/j.sbspro.2015.04.900.

Agyapong, D. and Tweneboah, G. (2023). The antecedents of circular economy financing and investment supply: The role of financial environment. Cleaner Environmental Systems 8. https://doi.org/10.1016/j.cesys.2022.100103.

Ali, Y., Uddin, A. and Petrillo, A. (2024). The impact of government support and organizational culture on sustainable performance: Unveiling the mediating role of circular economy and eco-innovation. Sustainable Futures 8. https://doi.org/10.1016/j.sftr.2024.100346.

Aranda-Usón, A., Portillo-Tarragona, P., Marín-Vinuesa, L. M. and Scarpellini, S. (2019). Financial resources for the circular economy: A perspective from businesses. Sustainability (Switzerland) 11(3). https://doi.org/10.3390/su11030888.

Araujo Galvão, G. D., De Nadae, J., Clemente, D. H., Chinen, G. and De Carvalho, M. M. (2018). Circular economy: Overview of barriers. Procedia CIRP 73: 79–85. https://doi.org/10.1016/j.procir.2018.04.011.

Bocken, N. M. P., de Pauw, I., Bakker, C. and van der Grinten, B. (2016). Product design and business model strategies for a circular economy. Journal of Industrial and Production Engineering 33(5): 308–320. https://doi.org/10.1080/21681015.2016.1172124.

Castro-Lopez, A., Iglesias, V. and Santos-Vijande, M. L. (2023). Organizational capabilities and institutional pressures in the adoption of circular economy. Journal of Business Research 161. https://doi.org/10.1016/j.jbusres.2023.113823.

Ghouri, A. M., Akhtar, P., Shahbaz, M. and Shabbir, H. (2019). Affective organizational commitment in global strategic partnerships: The role of individual-level microfoundations and social change. Technological Forecasting and Social Change 146: 320–330. https://doi.org/10.1016/j.techfore.2019.05.025.

Gonçalves, B. de S. M., de Carvalho, F. L. and Fiorini, P. de C. (2022). Circular economy and financial aspects: a systematic review of the literature. Sustainability (Switzerland) 14(5). https://doi.org/10.3390/su14053023.

Govindan, K. and Hasanagic, M. (2018). A systematic review on drivers, barriers, and practices towards circular economy: a supply chain perspective. International Journal of Production Research 56(1–2): 278–311. https://doi.org/10.1080/00207543.2017.1402141.

Karman, A., Viktor, P. and Ana Beatriz, L. de S. J. (2024). Circular economy practices as a shield for the long-term organisational and network resilience during crisis: Insights from an industrial symbiosis. Journal of Cleaner Production 142822.

Khan, S. A. R., Umar, M., Asadov, A., Tanveer, M. and Yu, Z. (2022). Technological revolution and circular economy practices: a mechanism of green economy. Sustainability (Switzerland) 14(8). https://doi.org/10.3390/su14084524.

Kirchherr, J. and Piscicelli, L. (2019). Towards an education for the circular economy (ECE): Five teaching principles and a case study. Resources, Conservation and Recycling 150.

Kirchherr, J., Piscicelli, L., Bour, R., Kostense-Smit, E., Muller, J., Huibrechtse-Truijens, A. et al. (2018). Barriers to the circular economy: evidence from the European Union (EU). Ecological Economics 150: 264–272. https://doi.org/10.1016/j.ecolecon.2018.04.028.

Korhonen, J., Honkasalo, A. and Seppälä, J. (2018). Circular economy: the concept and its limitations. Ecological Economics 143: 37–46. https://doi.org/10.1016/j.ecolecon.2017.06.041.

Lengnick-Hall, C. A., Beck, T. E. and Lengnick-Hall, M. L. (2011). Developing a capacity for organizational resilience through strategic human resource management. Human Resource Management Review 21(3): 243–255.

Lewandowski, M. (2016). Designing the business models for circular economy-towards the conceptual framework. In Sustainability (Switzerland) (Vol. 8, Issue 1). MDPI. https://doi.org/10.3390/su8010043.

Okoye, P. V. C. and Ezejiofor, R. A. (2013). The effect of human resources development on organizational productivity. International Journal of Academic Research in Business and Social Sciences 3(10). https://doi.org/10.6007/ijarbss/v3-i10/295.

Salvador, R., Barros, M. V., Freire, F., Halog, A., Piekarski, C. M. and De Francisco, A. C. (2021). Circular economy strategies on business modelling: Identifying the greatest influences. Journal of Cleaner Production 299. https://doi.org/10.1016/j.jclepro.2021.126918.

Sehnem, S., Chiappetta Jabbour, C. J., Farias Pereira, S. C. and de Sousa Jabbour, A. B. L. (2019). Improving sustainable supply chains performance through operational excellence: circular economy approach. Resources, Conservation and Recycling 149: 236–248. https://doi.org/10.1016/j.resconrec.2019.05.021.

Suárez-Eiroa, B., Fernández, E. and Méndez, G. (2021). Integration of the circular economy paradigm under the just and safe operating space narrative: Twelve operational principles based on circularity, sustainability and resilience. In Journal of Cleaner Production (Vol. 322). Elsevier Ltd. https://doi.org/10.1016/j.jclepro.2021.129071.

Suárez-Eiroa, B., Fernández, E., Méndez-Martínez, G. and Soto-Oñate, D. (2019). Operational principles of circular economy for sustainable development: Linking theory and practice. pp. 952–961. In Journal of Cleaner Production (Vol. 214). Elsevier Ltd. https://doi.org/10.1016/j.jclepro.2018.12.271.

Valbuena-Hernandez, J. P. and Ortiz-de-Mandojana, N. (2022). Encouraging corporate sustainability through effective strategic partnerships. Corporate Social Responsibility and Environmental Management 29(1): 124–134. https://doi.org/10.1002/csr.2188.

Wynn, M. and Jones, P. (2022). Digital technology deployment and the circular economy. Sustainability 14(15): 9077.

Yang, M., Chen, L., Wang, J., Msigwa, G., Osman, A. I., Fawzy, S. et al. (2023). Circular economy strategies for combating climate change and other environmental issues. pp. 55–80. In Environmental Chemistry Letters (Vol. 21, Issue 1). Springer Science and Business Media Deutschland GmbH. https://doi.org/10.1007/s10311-022-01499-6.

Chapter 4

Neuro-Fuzzy Modelling for Assessing Green Behaviour in Higher Education

*Adrian Vîlcu** and *Raluca Petronela Lăzărescu*

1. Introduction

The link between personality traits, emotional intelligence and green behaviour has been searched in the last decade. Individuals who develop skills and abilities regarding green behaviour tend to register high scores in the openness and agreeableness dimension of the Big Five personality Model. The researchers found strong correlations between emotional intelligence and specific personality traits – extraversion and autonomy (Sălceanu and Agapie, 2022). Personality traits and the level of education became essential factors for adopting responsible environmental behaviour together with proper actions to solve environmental issues (Wirdianti et al., 2019).

This chapter explores a complex theme at the intersection of psychology, education, and sustainability, focusing on the impact of personality traits and emotional intelligence on ecological behaviour, emphasizing the moderating role of educational level. The study employs an Adaptive Neuro-Fuzzy Inference System (ANFIS), a hybrid modelling method that uses a fuzzy system where the rule set is automatically generated by a neural network with supervised learning based on a back-propagation algorithm. This system evaluates the relationship between Conscientiousness (CO), Emotional Intelligence (EI), and Green Behavior

Gheorghe Asachi Technical University of Iasi, Faculty of Industrial Design and Business Management, 29 Prof. D. Mangeron Blvd, 700050, Iasi, Romania.
* Corresponding author: adrian.vilcu@academic.tuiasi.ro

DOI: 10.1201/9781003518990-4

(GB) among students at the "Gheorghe Asachi" Technical University of Iași. This approach contributes to understanding the mechanisms that determine the adoption of pro-environmental behaviour in the context of higher education. The ANFIS modelling allows for the analysis of the combined impact of CO and EI on GB, integrating educational level as a moderating variable.

The research involves 440 students divided into five educational levels: undergraduate years (1–4) and master's programs. Using validated questionnaires such as the Big Five Inventory for measuring personality traits, the PEBS scale for pro-environmental behaviour, and other tools for assessing EI, the study provides valuable insights into the dynamics of psychological and educational factors among individuals with higher education.

The study's results highlight that high EI is associated with increased involvement in pro-environmental actions due to enhanced Empathy, emotional control, and the ability to anticipate the long-term consequences of one's actions.

Thus, this chapter aims to provide an integrated perspective on the factors influencing students' ecological behaviour. By applying advanced modelling methods and offering recommendations for improving sustainability education, the research seeks to contribute to the specialized literature and support global efforts to promote more responsible and sustainable ecological behaviour among future generations.

2. Literature review

Emotional intelligence (EI) is a factor that significantly influences green behaviour, especially among the young population. Research has shown that higher levels of EI are associated with greater engagement in green behaviour (Lisboa et al., 2024; Zelenski and Desrochers, 2021). Adolescents with high EI are better able to manage their emotions, empathize with others, and understand the long-term consequences of their actions, making them more likely to engage in sustainable behaviours. Therefore, educational programs that enhance EI could play a crucial role in fostering green behaviour, particularly in younger populations.

Emotional awareness refers to the ability to recognize, understand, and differentiate one's own emotions and the emotions of others. It is a critical component of emotional intelligence and is vital in how individuals interact with the world, make decisions, and build relationships (Salovey and Mayer, 1990). Gross's research (Gross, 1998) is foundational in the study of emotion regulation, which includes emotional control as a core component.

Sustaining Green Behavior—Engaging in green behaviour often requires sustained effort and the ability to overcome setbacks or social pressures. Gross (1998), in his review of emotion regulation, highlights how emotional control contributes to persistence and consistency in goal-directed behaviours. This persistence is critical in green behaviour, where the benefits of actions like reducing energy consumption or adopting sustainable practices are often not immediately visible. Individuals with solid emotional control are likelier to remain committed to green behaviour despite potential challenges.

The correlation between emotional control and green behaviour is significant, with emotional control facilitating sustained engagement in environmentally responsible actions. Individuals who can manage their emotions effectively are better equipped to handle the challenges of green behaviour, including stress, social pressures, and the emotional burden of confronting environmental issues.

Research regarding the influence of demographic factors on green behaviour indicates that age, gender, education and income influence environmental attitudes and actions. Individuals with higher education and income tend to act in a green direction, developing behaviours in order to protect and save the environment, especially on climate issues (Yang and Arhonditsis, 2022).

The green behaviour regarding recycling, reducing waste and reusing resources is based on attitudes, perceived effectiveness and demographic factors (gender, age, education). Research demonstrated that higher levels of formal education increase individuals′ environmentally conscious behaviour and create attitudes toward green behaviour because of the environmental acquisition of information (Wang et al., 2022). The awareness regarding the climate change causes and self-efficacy of green behaviour was recorded in people with a high degree of education (Hoffmann and Muttarak, 2020)

While numerous factors influence green behaviour, personality traits, particularly those within the Big Five model, have been shown to play a significant role in determining how likely an individual is to engage in such behaviours (Jach et al., 2023; Kesenheimer and Greitemeyer, 2021). Extraversion is one of the Big Five personality traits, typically characterized by high levels of energy, enthusiasm, and a propensity for seeking out social interactions (Lukaszewski and von Rueden, 2015) and a dimension associated with sociability, assertiveness, and a preference for stimulation—has shown a complex and sometimes contradictory relationship with green behaviour (Eysenck, 1983). Given these characteristics, one might intuitively expect extroverts to be more engaged in community-oriented green behaviours, such as participating in environmental activism or community clean-up events.

Extraverts may be more likely to engage in green behaviour when it offers opportunities for social recognition or status enhancement. For instance, volunteering for high-profile environmental causes or adopting visible green practices like driving an electric vehicle may appeal to extroverts who enjoy being seen as leaders or trendsetters within their social circles. This aligns with the "green consumerism" concept, where environmentally conscious consumption is partly driven by a desire for social approval (Moisander and Pesonen, 2002).

Agreeableness—a dimension characterized by Empathy (Graziano et al., 2007), cooperation, and a concern for social harmony (Ashton et al., 2014)—has been strongly associated with pro-environmental. Green actions often require individuals to consider the broader social and ecological impacts of their behaviour, and those who are high in agreeableness are more likely to engage in such behaviours out of a sense of moral responsibility and concern for others. For instance, a meta-analysis by Soutter et al. (2020) found that agreeableness was one of the strongest predictors of green behaviour among the Big Five personality traits.

Empathy, a core component of agreeableness, enables individuals to understand and share the feelings of others, including those affected by environmental degradation. This emotional connection can drive green actions, as agreeable individuals feel a moral obligation to protect the environment for the sake of others. Research by Kesenheimer and Greitemeyer (2021) supports this view, showing that individuals high in agreeableness are more likely to engage in behaviours that reduce their environmental impact because they empathize with the victims of environmental harm. Moreover, agreeableness is associated with a strong sense of moral responsibility, further driving green behaviour (Graziano et al., 2007).

A study by Jach et al. (2023) found that agreeableness was positively correlated with participation in collective environmental actions, such as joining environmental organizations or attending protests. The researchers suggest that agreeable individuals are motivated by a desire to work with others toward a common goal, mainly when that goal involves protecting the environment.

Conscientiousness is one of the Big Five personality traits, reflecting a tendency toward carefulness, diligence, and reliability. Numerous studies have demonstrated a positive correlation between Conscientiousness and various forms of green behaviour. Individuals high in Conscientiousness are more likely to engage in behaviours that require careful planning and persistence, such as energy conservation (Hirsh, 2010), recycling (Swami et al., 2010), and sustainable consumption. This ability to delay gratification and focus on long-term outcomes is a hallmark of Conscientiousness and is crucial in promoting green behaviour. Hirsh and Dolderman (Hirsh and Dolderman, 2007) emphasize that conscientious individuals are likelier to engage in behaviours that require self-control, such as reducing consumption and avoiding impulsive purchases.

Clear guidelines and structured programs that help conscientious individuals integrate green behaviour into their routines can enhance their engagement. For instance, offering step-by-step plans for reducing energy consumption or detailed instructions for recycling can appeal to conscientious individuals' orderly and organized nature, making it easier for them to adopt and maintain these behaviours.

Neuroticism is one of the Big Five personality traits. It is associated with a tendency to experience negative emotions such as anxiety (Lahey, 2009), fear (McCrae and Costa, 1991), sadness (Watson and Clark, 1984), and irritability (Lahey, 2009). Individuals high in neuroticism are often emotionally reactive and may struggle with stress management (Suls and Martin, 2005). They are more likely to perceive situations as threatening or overwhelming and are prone to worry and self-doubt. Given these characteristics, one might expect neuroticism to have a mixed impact on green behaviour. On the one hand, the heightened sensitivity to threats, including environmental risks, could motivate individuals high in neuroticism to engage in behaviours to mitigate these threats.

The way environmental issues are framed can significantly influence how neurotic individuals respond.

McCrae and Costa (McCrae and Costa, 1997) define openness as a trait characterized by imagination, curiosity, creativity, and a preference for variety and intellectual experiences, and John and Srivastava (John and Srivastava, 1999) describe openness as a trait associated with aesthetic sensitivity, attentiveness to inner feelings, preference for variety, and intellectual curiosity. These characteristics have significant implications for green behaviour, as they can influence how individuals perceive and respond to environmental issues. Their curiosity drives them to explore new ideas and perspectives, which can lead to a greater understanding of the complexities of environmental problems and the importance of sustainable practices.

Openness is also associated with creativity and a willingness to experiment with new ideas and approaches. This trait can lead to innovative solutions to environmental challenges and a greater willingness to adopt unconventional or experimental sustainable practices. Creative Problem Solving – individuals high in openness are often skilled at creative problem solving, which can be applied to environmental issues. They may be more likely to experiment with new technologies (Hirsh, 2010), such as renewable energy sources (Markowitz et al., 2012), or to adopt innovative approaches to reduce waste and conserve resources (Milfont and Sibley, 2012). This creative approach to problem-solving is essential for addressing complex and multifaceted environmental challenges.

3. Research methodology

The study's target population is students. The sample is considered uniform weight. The participants are 440 students enrolled in degree and master programmes studying at Gheorghe Asachi Technical University of Iaşi, Romania. The participants were informed regarding the questionnaire contents and the scale and encouraged to use the feedback in case of misunderstandings. Participation was voluntary. The response was online through a form that was distributed to the student group. The data collection period was May–June 2024, at the end of the academic year.

The authors used an adapted emotional intelligence questionnaire (Whetten and Cameron, 2011), including items with three types of responses referring to specific reactions in different situations. The questionnaire includes four dimensions: Emotional Awareness, the ability to recognize emotions at a personal level and to integrate them with others' emotions in a specific context of the communication process; Emotional Control (Balance), the ability to have a proper response in stressful communication situations, Emotional Diagnosis (Empathy): the ability to understand others' feelings and act in a manner to show openness in the communication process, Emotional Response: the feedback at an emotional level that a person may offer in communication process.

The instrument used by authors to establish correlations between personality factors and green behaviour was the Big Five Inventory (BFI), a self-report inventory designed to measure the Big Five dimensions: extraversion, agreeableness, conscientiousness, neuroticism, and openness. The instrument was developed by (John et al., 1991).

In order to measure green behaviour, the authors adapted the Pro-Environmental Behaviours Scale (PEBS) (Markle, 2013; Menardo et al., 2019) and the Pro-environmental Attitudes Questionnaire (PEAQ) (Felonneau and Becker, 2008) considered by previous researchers being significantly essential to validate the results. The dimensions are economics, resource usage, protection and preservation of the natural environment, culture and individual mentality toward green behaviour, education – the information level, organizations' social and green strategy, and governmental measures. The questionnaires used a Likert scale: 1 - strongly disagree, 2 - disagree, 3 - neutral, 4 - agree, 5 - strongly agree.

The questionnaire, which was applied online via the Google Forms platform, was answered by 440 students, with the following distribution: 40% male and 60% female. Their academic levels were distributed as follows: Bachelor's students - Year 1 (26%), Year 2 (32%), Year 3 (19%), Year 4 (9%), and Master's Years 5+6 (14%). Regarding geographical distribution, 22% were from the Republic of Moldova and 78% from Romania. Their fields of study were 66% in Engineering and Management, 10% in Machine Building, 16% in Electrical Engineering, Electrotechnics, and Applied Informatics, and 8% in Electronics, all from the "Gheorghe Asachi" Technical University of Iași, Romania.

The dimensions of the research model are:

Emotional Intelligence dimension [EI]: measured through a 12-item questionnaire grouped into four dimensions: Emotional Awareness (3 items), Emotional Control (3 items), Empathy (3 items), and Emotional Response (3 items). The weight of all four dimensions is considered equal. Therefore, the relationship defining the EI variable is:

$$[\mathrm{EI}] = \frac{\sum_{i=1}^{12} \mathrm{Item}_i}{12}$$

The [EI] dimension takes values in the interval [0…10], with the minimum value characterizing a personality with a shallow level of emotional intelligence and the maximum value characterizing a person capable of identifying and understanding emotions, controlling their feelings and expressions, demonstrating high Empathy by experiencing the same emotions as their interlocutor, and, lastly, being able to transform one emotion into another.

Conscientiousness dimension [CO]: one of the five dimensions in the Big-Five questionnaire for personality factors (Agreeableness, Extraversion, Conscientiousness, Neuroticism, and Openness). This [CO] dimension includes nine items (Table 1). This dimension was chosen because Conscientiousness reflects an individual's tendency to be organized, responsible, perseverant, and goal-oriented. Critical characteristics of [CO] that correlate with a high degree of [GB] (Green Behaviour) are: *order* that reflects a high level of self-organization, directly influencing environmental sustainability and rational resource consumption, *responsibility* - involves strong adherence to norms, including environmental norms, *self-discipline* - indicates

Table 1. Items for the [CO] dimension.

Item	Reversed	Item Description
1	No	I am a person who is attentive to details.
2	Yes	I am a person who might be considered careless.
3	No	I am a person others can rely on to complete a task.
4	Yes	I am a person who tends to be disorganized.
5	Yes	I am a person who tends to be lazy.
6	No	I am a persevering person. I keep going until I achieve my goal.
7	No	I am a person who does things efficiently.
8	No	I am a person who makes plans and follows them strictly.
9	Yes	I am a person who is easily distracted.

effective planning and reduced impulsive resource consumption, *achievement orientation* - supports the sustainable use of environmental resources.

For the [CO] dimension, the calculation formula is

$$[\mathrm{CO}] = \frac{\sum_{i=1}^{9} \mathrm{Item}_i}{9}$$

The value range is [0…5], where the minimum value represents a low level of Conscientiousness, while the maximum value indicates a very high level of this trait.

Education Level [EL]: the moderating variable which takes values from 1 to 5 (1–4: Bachelor's, 5: Master's).

Green Behaviour dimension [GB]: the output variable comprising 50 items grouped into nine dimensions (Table 2).

Table 2. The structure of the [GB] dimension.

Dimension	Number of Items
Pro-environment Economic Dimension	3
Resource Consumption	8
Nature Conservation Dimension	8
Pro-environment Cultural Dimension	5
Individual Mentality Dimension	6
Pro-environment Education Dimension	6
Social Dimension	8
Pro-environment Corporate Dimension	3
Government Pro-environment Measures	3

The calculation formula is:

$$[GB] = \frac{\sum_{i=1}^{9} \left(\frac{\sum_{j=1}^{n_i} Item_j}{n_i} \right)}{9}$$

Where n_i is the number of items in the Dim_i dimension.

The [GB] variable takes values in the interval [0…5]. A value of 0 characterizes individuals with an attitude, position, and actions against environmental protection. In contrast, a value of 5 characterizes individuals who adopt behaviors, policies, and habits that support environmental protection, respect sustainability principles, reduce negative impacts on nature, and promote a harmonious relationship between humans and natural ecosystems.

The proposed research model is a multivariable type [GB]/[EL] = f([EI],[CO]) (Figure 1) with a black-box structure. This simulates the system by highlighting the connections between input variables [EI], [CO], and the output variable [GB] and observing the effect of the moderating variable [EL] on the internal structure of the Adaptive Neuro-Fuzzy Inference System (ANFIS).

In Figure 1, the variable [Education level] = [EL] is considered a moderating variable that moderates the relationship between the input variables [Emotional Intelligence] = [EI], [Personality Factors] = [PF], and the output variable [Green Behavior] = [GB].

Respondents were grouped into five levels of the variable [EL], as shown in Table 3.

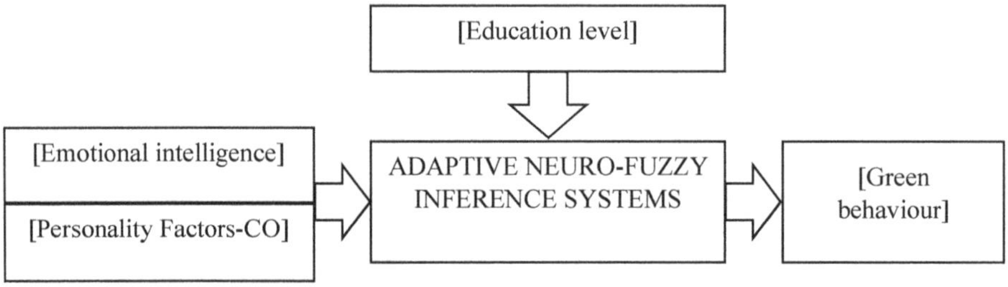

Figure 1. The dimensions of the research model and the ANFIS structure.

Table 3. Grouping of responses for training the ANFIS modelling structures.

Education Level	**Year of Study**	**Number of Respondents/ Training Patterns (ANFIS)**	**ANFIS Coding**
1	Undergraduate Year 1	115	[EL1]
2	Undergraduate Year 2	140	[EL2]
3	Undergraduate Year 3	83	[EL3]
4	Undergraduate Year 4	40	[EL4]
5	Master's Years 1 and 2	62	[EL5]

The simulation of the research model [GB]/[EL] = f([EI], [CO]), where [EL] is considered a moderating variable, was conducted using a Fuzzy System with a Takagi–Sugeno inference algorithm. It involved two inputs [PF], [EI], and one output [GB], where the rule system was automatically generated by a Neural Network using supervised learning with a backpropagation training algorithm. The fuzzification and defuzzification functions are Gaussian in three levels. The programming environment used was MATLAB 2016.

Five ANFIS systems were developed, one for each education level, trained with distinct datasets resulting from clustering student responses from undergraduate Years 1–4 and Master's Years 1 and 2.

4. Results and discussions

The behavior of the five ANFIS systems is presented in Table 4.

At Education Level 1, the ANFIS generates a maximum level for [GB] at a maximum level of the variable [CO] and an average value of the variable [EI] (Figure 2.a). The response surface represents the minima of the dependent variable [GB] for low levels of [CO] and extreme values of the variable [EI], a behavior also observed in (Lisboa et al., 2024). The model suggests an almost linear dependency between the dimension [GB] and [EI] (Figure 2.b), at a high level of the dimension [GB], in the range (3.5…4.3). The model has a plateau in the interval (2…4) for the dimension [CO] (Figure 2.c) at a level of 3.8 for the dimension [GB], while in the interval (4…5), the variable [GB] shows an abrupt increase. Thus, for education level 1, the results provided by the ANFIS trained with 115 input-output patterns lead to the conclusion that the dimension [GB] is almost linearly dependent on the dimension [EI] and exponentially dependent on the dimension [CO].

The ANFIS for Education level 2, trained with 140 input-output patterns, determines a model different from Education level 1 for the same research model [GB] = f([EI], [CO]) (Figure 3.a). Thus, the intervals for the variables are more compact (3.75…4.05) for the dimension [EI] and [3.9…4.15] for the dimension [CO]. A two-tier behavior can be observed in the dependence of the variable [GB] on [EI], separated by a median value of 4 for [EI] (Figure 3.b). A near-maximum level ([GB] = 5) is reached for a larger overlap interval of the input parameters [EI] $\in$ (4…8) and [CO] $\in$ (3…5). The variable [GB] is observed to be an increasing, almost linear function of the variable [CO] (Figure 3.c).

The ANFIS for Education level 3 reveals a different dynamic behavior across three zones of the variable [EI] (Figure 4.b) and three zones of the variable [CO] (Figure 4.c), influencing the variable [GB]. High values of [GB] between (3.5…4.5) are achieved in the extreme zones, 1 and 3, of [EI] and [CO] (Figure 4.a), while in the intermediate zones, the level of the variable [GB] is lower, reaching as low as 3.

The ANFIS for Education level 4 (3 years more on average than respondents in level 1) shows behavior similar to the ANFIS trained on the data of first-year respondents but with a smoother surface evolution (Figure 5.a). Maximum values of [GB] are obtained at the intersection of high values of [CO] (in the range (3…5)) and relatively low values of [EI] (below 5, out of a maximum of 10). [GB] is a

Table 4. The ANFIS behavior.

<table>
<tr><th>Ed. level</th><th>The relationship between [GB] and [EI], [CO]</th><th>The relationship between [GB] and [EI]</th><th>The relationship between [GB] and [CO]</th></tr>
<tr><td>[EL1]</td><td>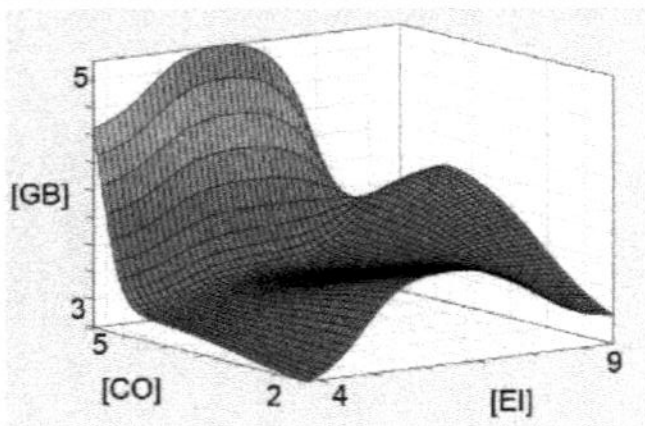

Figure 2.a [GB] = f([CO],[EI])</td><td>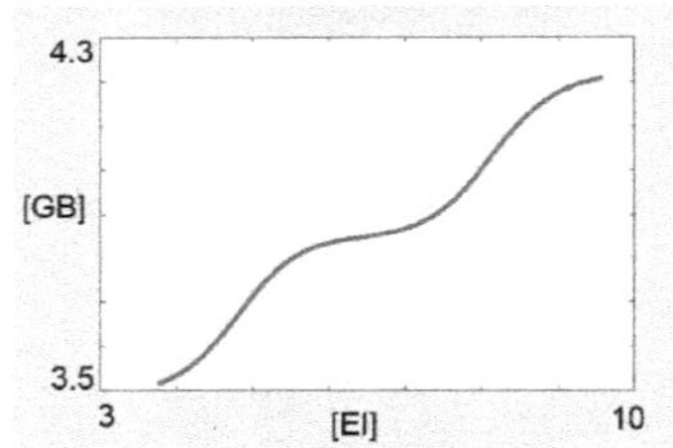

Figure 2.b [GB] = f([EI])</td><td>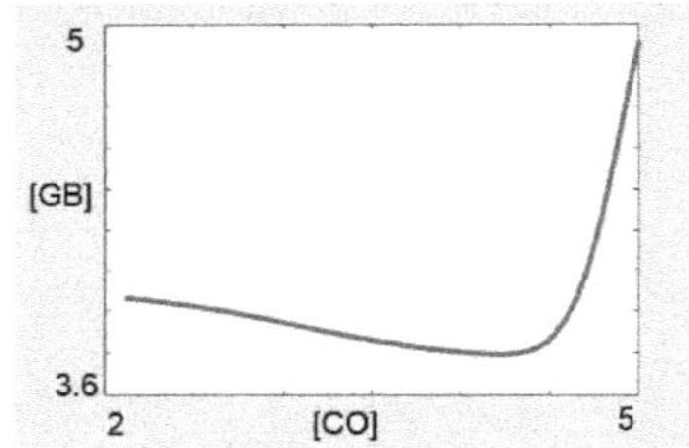

Figure 2.c [GB] = f([CO])</td></tr>
<tr><td>[EL2]</td><td>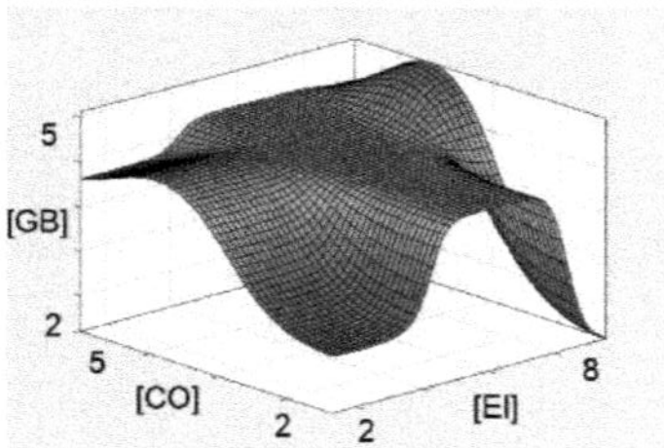

Figure 3.a [GB]=f([CO],[EI])</td><td>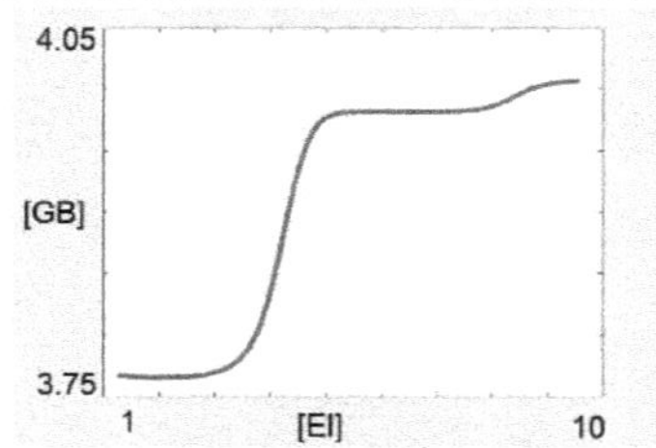

Figure 3.b [GB]=f([EI])</td><td>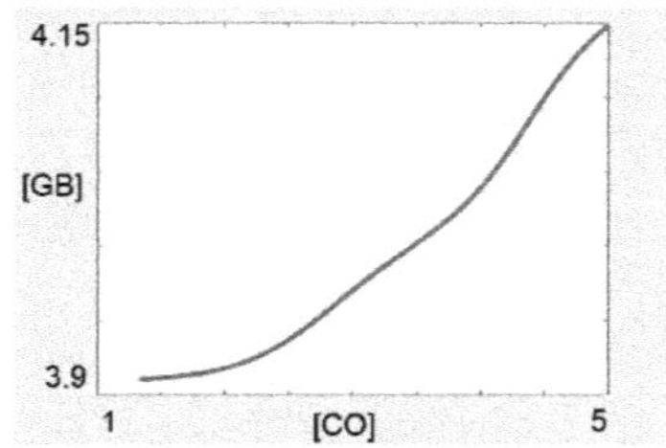

Figure 3.c [GB]=f([CO])</td></tr>
<tr><td>[EL3]</td><td>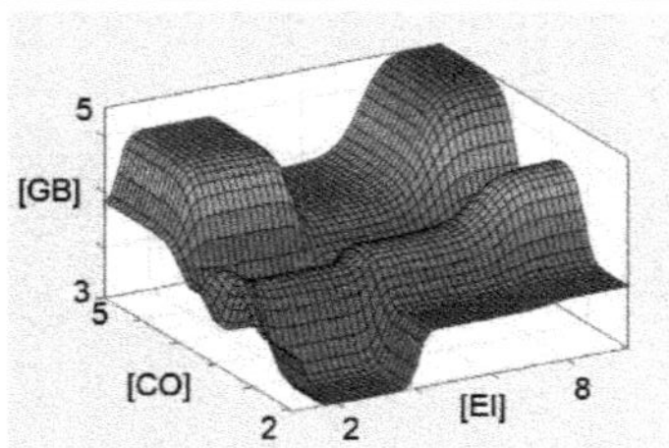

Figure 4.a [GB]=f([CO],[EI])</td><td>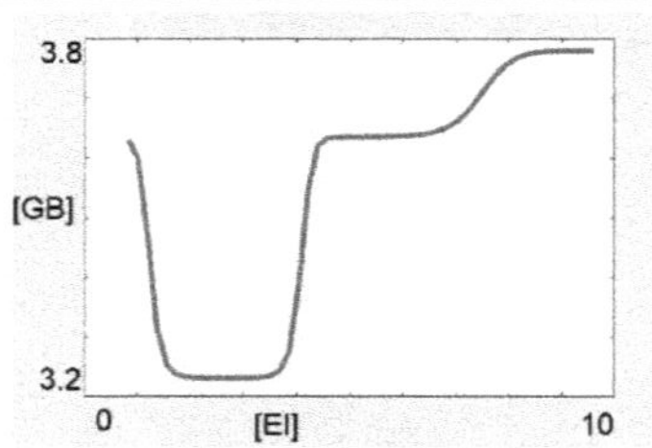

Figure 4.b [GB]=f([EI])</td><td>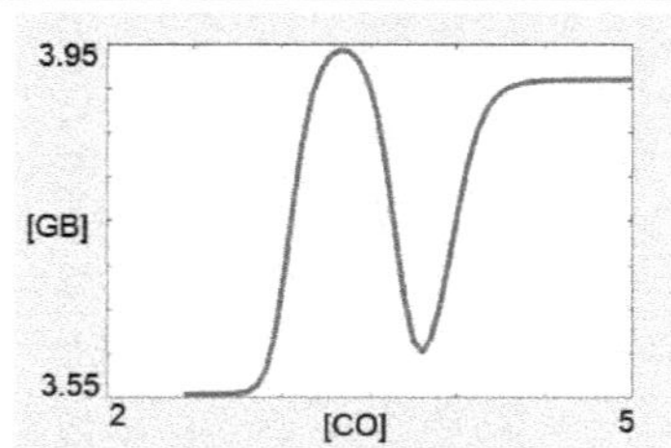

Figure 4.c [GB]=f([CO])</td></tr>
</table>

<table>
<tr>
<td>[EL4]</td>
<td>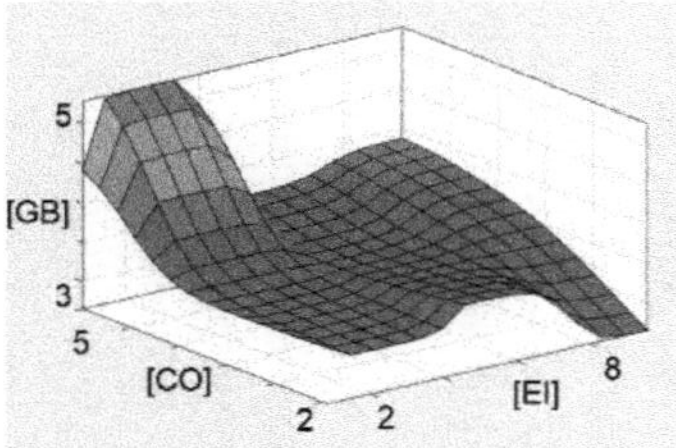

Figure 5.a [GB]=f([CO],[EI])</td>
<td>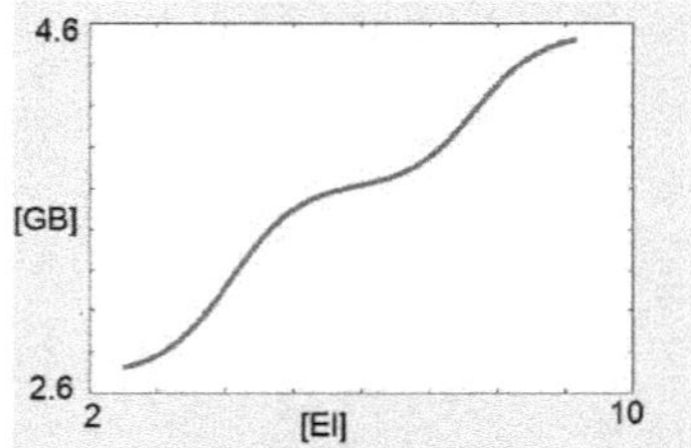

Figure 5.b [GB]=f([EI])</td>
<td>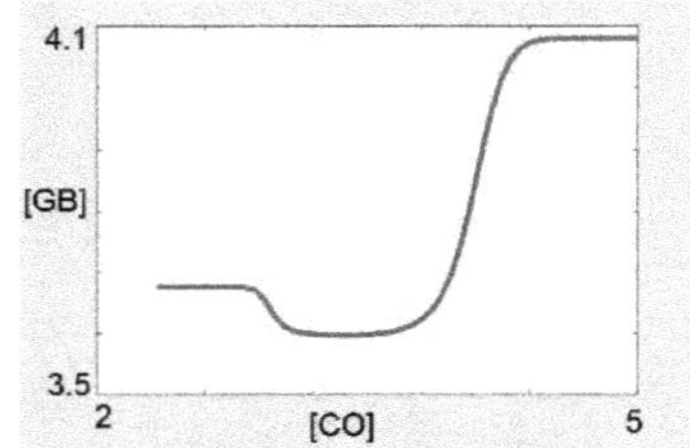

Figure 5.c [GB]=f([CO])</td>
</tr>
<tr>
<td>[EL5]</td>
<td>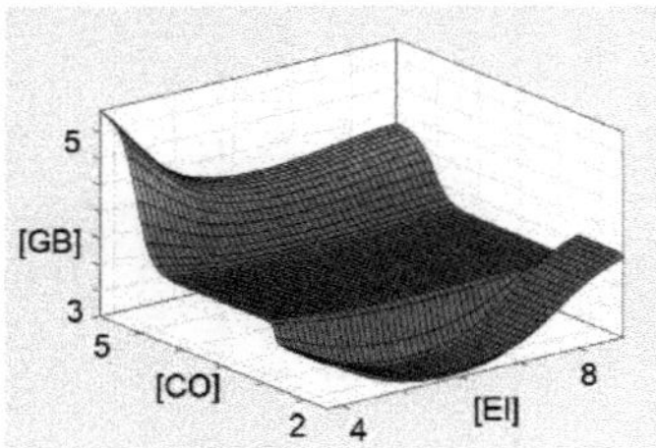

Figure 6.a [GB]=f([CO],[EI])</td>
<td>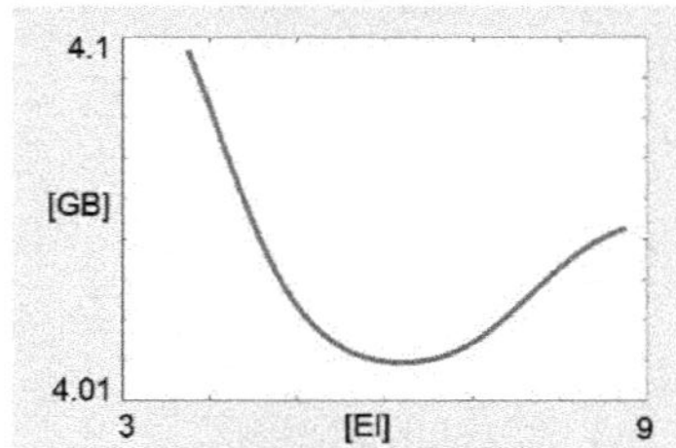

Figure 6.b [GB]=f([EI])</td>
<td>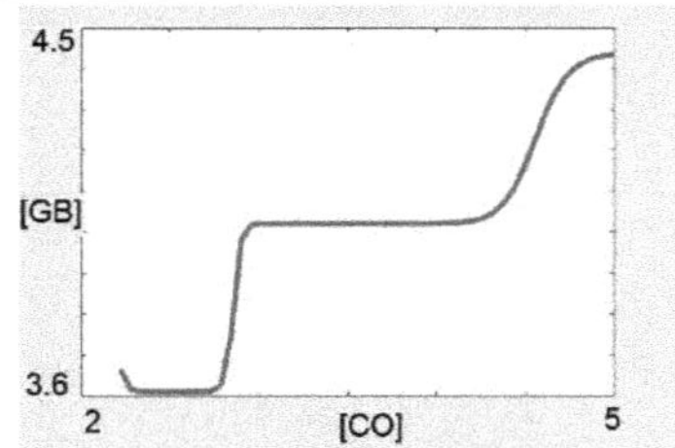

Figure 6.c [GB]=f([CO])</td>
</tr>
</table>

linear function of [EI] (Figure 5.b) and an exponential function with saturation at a maximum for [CO] values (Figure 5.c) greater than 3.

The ANFIS for Education level 5 (master's students) presents a different result than the first four structures. [GB] shows a parabolic graph with the variable [EI] (Figure 6.b) and a stepped function with the variable [CO] (Figure 6.c), with a larger intermediate plateau in the range (3…4). The maximum values of the variable [GB] are obtained for high values of [CO] across the entire range of values of the variable [EI] (Figure 6.a). The variation of [GB] with the variable [EI] is very tight in the interval (4.01… 4.1).

5. Conclusions

A primary conclusion of this study is that individuals with a high level of Conscientiousness tend to adopt eco-friendly behaviors, conserve resources, and use them rationally because they are more sensitive to the consequences of their actions on the environment. This aspect is also correlated with high emotional intelligence ([EI]). The research suggests that the self-discipline and responsibility associated with Conscientiousness motivate students to plan and implement strategies to reduce their ecological impact regardless of their education level.

The ANFIS systems generated for the five education levels indicate that students with high Conscientiousness report more active ecological behavior in their personal and social lives. In general, individuals with a high level of education (undergraduate and master's degrees) are more responsible and more likely to assess the long-term impact of their actions on the environment.

Another conclusion is that individuals who demonstrate careful planning and caution often make sustainable decisions, which are characteristics of [CO].

The positive correlation between the dimensions [GB] and [CO]—also observed in this research—is based on how personality traits motivate individuals with higher education to be more engaged and responsible in adopting ecological behaviors. This is supported by studies showing that conscientious individuals tend to prioritize environmental protection actions as part of their moral and personal commitment.

[EI] contributes to sustainability by influencing pro-environment behaviors at both individual and collective levels. This research highlights the importance of integrating [EI] into education and leadership strategies focused on sustainability for a more significant ecological impact.

The current research also evaluated the role of the education level as a moderating variable in the relationship between [EI], [CO], and [GB]. The study examined the positive effect of education, which can strengthen the relationship between [CO] and sustainable behaviors—because educated individuals tend to understand better the long-term importance of their ecological actions—and the negative or limiting moderating effect of education on environmental sustainability.

The study results show that the moderating effect of the [EL] variable on the [EI] + [CO] → [GB] system varies, alternating between positive and negative effects across consecutive increasing education levels. The conclusion is that the lack of pro-environment education in higher education is a real issue that has implications for students' awareness and future actions in sustainability. This limits the preparation of future generations to address ecological problems and implement sustainable solutions.

Recommendations include integrating sustainability into curricula and fostering EI and CO as part of educational strategies. By leveraging emotional intelligence and Conscientiousness, educators and policymakers can better prepare students to tackle environmental challenges, ensuring that sustainability becomes an integral part of their professional and personal decision-making processes.

References

Ashton, M. C., Lee, K. and de Vries, R. E. (2014). The HEXACO honesty-humility, agreeableness, and emotionality factors: a review of research and theory. Personality and Social Psychology Review 18(2): 139–152. https://doi.org/10.1177/1088868314523838.

Eysenck, H. J. (1983). 2 - Psychophysiology and personality: extraversion, neuroticism and psychoticism. Physiological Correlates of Human Behaviour 3: 13–30. https://doi.org/10.1016/B978-0-12-273903-3.50007-9.

Felonneau, M. L. and Becker, M. (2008). Pro-environmental attitudes and behaviour: Revealing perceived social desirability. Revue internationale de psychologie sociale 2008/4 Tome 21, 25–53. Presses univ. de Grenoble Presses univ. de Grenoble Powered by TCPDF (www.tcpdf.org) 2008/4, ISSN 0992-986X ISBN 9782706115264.

Graziano, W. G., Habashi, M. M., Sheese, B. and Tobin, R. M. (2007). Agreeableness, empathy, and helping: A person × situation perspective. Journal of Personality and Social Psychology 93(4): 583–599. https://doi.org/10.1037/0022-3514.93.4.583.

Gross, J. J. (1998). The emerging field of emotion regulation: An integrative review. Review of General Psychology pp. 271–299. https://doi.org/10.1037/1089-2680.2.3.2.

Hirsh, J. B. (2010). Personality and environmental concern: The role of openness to experience and Conscientiousness. Journal of Environmental Psychology 30(2): 245–248. https://doi.org/10.1016/j.jenvp.2010.01.004.

Hirsh, J. B. and Dolderman, D. (2007). Personality predictors of consumerism and environmentalism: A preliminary study. Personality and Individual Differences 43(6): 1583–1593. https://doi.org/10.1016/j.paid.2007.04.015.

Hoffmann, R. and Muttarak, R. (2020). Greening through schooling: understanding the link between education and pro-environmental behaviour in the Philippines. Environmental Research Letters 15(1). https://doi.org/10.1088/1748-9326/ab5ea0.

Jach, H. K., Bardach, L. and Murayama, K. (2023). How personality matters for education research. Educational Psychology Review. https://doi.org/10.1007/s10648-023-09807-4.

John, O. P. and Srivastava, S. (1999). The Big Five trait taxonomy: History, measurement, and theoretical perspectives. Handbook of Personality: Theory and Research, New York: Guilford Press, 102–138. https://jenni.uchicago.edu/econ-psych-traits/John_Srivastava_1995_big5.pdf.

John, O. P., Donahue, E. M. and Kentle, R. L. (1991). The Big Five Inventory - Versions 4a and 54. Berkeley, CA: University of California, Berkeley, Institute of Personality and Social Research. https://sjdm.org/dmidi/Big_Five_Inventory.html.

Kesenheimer, J. S. and Greitemeyer, T. (2021). Going Green (and Not Being Just More Pro-Social): Do Attitude and Personality Specifically Influence Pro-Environmental Behavior? Sustainability. https://doi.org/10.3390/su13063560.

Lahey, B. (2009). Public health significance of neuroticism. American Psychologist 64(4): 241–256. https://doi.org/10.1037/a0015309.

Lisboa, P., Gomez-Roman, C., Guntin, L. and Monteiro, A. P. (2024). Pro-environmental behaviour, personality and emotional intelligence in adolescents: a systematic review. Frontiers in Psychology https://doi.org/10.3389/fpsyg.2024.1323098.

Lukaszewski, A. W. and von Rueden, C. R. (2015). The extraversion continuum in evolutionary perspective: A review of recent theory and evidence. Personality and Individual Differences 77: 186–192. http://dx.doi.org/10.1016/j.paid.2015.01.005.

Markle, G. L. (2013). Pro-environmental behaviour: Does it matter how it's measured? Development and validation of the pro-environmental behaviour scale (PEBS). Human Ecology 41: 905–914. https://doi.org/10.1007/s10745-013-9614-8.

Markowitz, E. M., Goldberg, L. R., Ashton, M. C. and Lee, K. (2012). Profiling the 'pro-environmental individual': A personality perspective. Journal of Personality 80(1): 81–111. https://doi.org/10.1111/j.1467-6494.2011.00721.x.

McCrae, R. R. and Costa, P. T. (1997). Personality trait structure as a human universal. American Psychologist 52(5): 509–516. https://doi.org/10.1037/0003-066X.52.5.509.

McCrae, R. and Costa, P. (1991). Adding liebe und Arbeit: The full five-factor model and well-being. Personality and Social Psychology Bulletin 17(2): 227–232. https://doi.org/10.1177/0146167291017002.

Menardo, E., Brondino, M. and Pasini, M. (2019). Adaptation and psychometric properties of the Italian version of the Pro-Environmental Behaviours Scale (PEBS). Environment, Development and Sustainability 6907–6930. https://doi.org/10.1007/s10668-019-00520-3.

Milfont, T. and Sibley, C. (2012). The big five personality traits and environmental engagement: Associations at the individual and societal level. Journal of Environmental Psychology 32(2): 187–195. https://doi.org/10.1016/j.jenvp.2011.12.006.

Moisander, J. and Pesonen, S. (2002). Narratives of sustainable ways of living: constructing the self and the other as a green consumer. Management Decision 40(4): 329–342. https://doi.org/10.1108/00251740210426321.

Sălceanu, C. and Agapie, O. M. (2022). Emotional intelligence and personality traits in higher education. Technium Social Sciences Journal 33: 416–429. https://doi.org/10.47577/tssj.v33i1.6965.

Salovey, P. and Mayer, J. D. (1990). Emotional intelligence. Imagination, Cognition and Personality 9(3): 185–211. https://doi.org/10.2190/DUGG-P24E-52WK-6CDG.

Soutter, A. B., Bates, T. C. and Mottus, R. (2020). Big Five and HEXACO personality traits, proenvironmental attitudes, and behaviors: a meta-analysis. Perspectives on Psychological Science 15(4): 913–941. https://doi.org/10.1111/jopy.12576.

Suls, J. and Martin, R. (2005). The daily life of the garden-variety neurotic: Reactivity, stressor exposure, mood spillover, and maladaptive coping. Journal of Personality 73(6): 1485–1510. https://doi.org/10.1111/j.1467-6494.2005.00356.x.

Swami, V., Chamorro-Premuzic, T., Snelgar, R. and Furnham, A. (2010). Personality, individual differences, and demographic antecedents of self-reported household waste management behaviours. Journal of Environmental Psychology 31(1): 21–26. https://doi.org/10.1016/j.jenvp.2010.08.001.

Wang, Q., Niu, G. and Gan, X. (2022). Green returns to education: Does education affect pro-environmental attitudes and behaviours in China? PLOS ONE. https://doi.org/10.1371/journal.pone.0263383.

Watson, D. and Clark, L. (1984). Negative affectivity: The disposition to experience aversive emotional states. Psychological Bulletin 96(3): 465–490. https://doi.org/10.1037/0033-2909.96.3.465.

Whetten, D. A. and Cameron, K. S. (2011). Developing management skills. pp. 71–75. Pearson Education, Prentice Hall, ISBN 10: 0-13-612100-4 ISBN 13: 978-0-13-612100-8 https://faculty.ksu.edu.sa/sites/default/files/developing_management_skills-8th_edition.pdf.

Wirdianti, N., Komala, R. and Miarsyah, M. (2019). Naturalist intelligence and personality: An understanding of students' responsible environmental behaviour. Jurnal Pendidikan Biologi Indonesia 229–236. https://doi.org/10.22219/jpbi.v5i2.7193.

Yang, C. and Arhonditsis, G. B. (2022). What are the primary covariates of environmental attitudes and behaviours in Canada? A national-scale analysis of socioeconomic, political, and demographic factors. Journal of Ecological Informatics. https://doi.org/10.1016/j.ecoinf.2022.101661.

Zelenski, J. M. and Desrochers, J. E. (2021). Can positive and self-transcendent emotions promote pro-environmental behaviour? Current Opinion in Psychology 31–35. https://doi.org/10.1016/j.copsyc.2021.02.009.

Chapter 5

Enhanced Model for Evaluating the Information Systems Success from the Sustainability Management Perspective

Pohonțu-Dragomir Ștefana-Cătălina,[1,*] *Herghiligiu Ionuț-Viorel,*[1,2] *Pislaru Marius*[1] and *Lăzărescu Raluca Petronela*[1]

1. Introduction

Evaluating information systems (IS) success from the sustainability management perspective is an emerging field that considers how information technology (IT) systems contribute to environmental, social, and economic sustainability goals. In the following, we present a bibliography of sources that explore the evaluation of IS success within the context of sustainability management.

Melville's paper discusses how information systems can be aligned with sustainability goals, focusing on innovations that support environmental sustainability (Melville, 2010). Hevner and Chatterjee, while not specifically focused on sustainability, provide methods for evaluating the success of IS design, which can be applied in a sustainability context (Hevner and Chatterjee, 2010).

Bharati and Chaudhury made a comprehensive literature review on the intersection of IS and sustainability, with an emphasis on frameworks and methodologies for evaluating IS success in sustainability contexts (Bharati and Chaudhury, 2015); Fitzgerald and Russo paper focuses on strategic approaches for managing IS within

[1] Gheorghe Asachi Technical University of Iasi, Faculty of Industrial Design and Business Management.
[2] Academy of Romanian Scientists, Ilfov 3, 050044 Bucharest, Romania.
* Corresponding author: stefana-catalina.pohontu-dragomir@student.tuiasi.ro

DOI: 10.1201/9781003518990-5

the context of sustainability and identifies challenges in measuring their success (Fitzgerald and Russo, 2012).

United Nations Global Compact and Accenture report explores how digital transformation and information systems can be leveraged to support the United Nations Sustainable Development Goals (SDGs) and evaluates their role in sustainability management (United Nations Global Compact & Accenture, 2019); World Economic Forum report examines how digital technologies and information systems can advance sustainability goals and provides metrics for assessing their success in this context (World Economic Forum, 2020).

These resources provide a variety of perspectives on evaluating the success of information systems in the context of sustainability management. They address multiple dimensions, including environmental, social, and economic sustainability, and provide frameworks, methodologies, and case studies that can help organizations assess how effectively their IS contributes to sustainability objectives.

2. Literature review

Evaluating information systems success from a sustainability management perspective is a growing area of interest in both academia and practice. In the following, we present some notable authors and works that explore this intersection of evaluating information systems success from the perspective of sustainability management. These studies explore how IS can be leveraged to promote environmental, social, and economic sustainability within organizations.

Melville investigates how IS innovations can foster environmental sustainability by reducing environmental impacts and supporting green practices. He explores how businesses can use IS to not only comply with environmental regulations but also to improve their sustainability performance. He examines the role of IS innovations in reducing environmental impacts and supporting sustainable business practices. It provides a framework for assessing the relationship between IS and environmental sustainability (Melville, 2010).

Bélanger and Crossler article, while primarily concerned with privacy, touches upon sustainability as it relates to ethical data management practices. It highlights the ethical implications of IS usage in sustainability practices, especially in terms of data privacy and security. Although primarily focused on privacy, this paper also touches upon sustainability from an ethical data management standpoint, a critical element of responsible information systems use (Bélanger and Crossler, 2011).

DeLone and M. M. B. paper discusses how IS can enhance sustainability by supporting decision-making processes that balance environmental, economic, and social considerations. The paper explores the broader implications of information systems for sustainability management, emphasizing long-term value creation through sustainable business practices. It focuses on the intersection of business value and sustainability, highlighting how IS can support long-term sustainable outcomes for organizations (DeLone and M. M. B., 2012).

Hedman and Valentin examine how organizations in different industries use IS to address sustainability challenges. They provide empirical evidence on how IS help organizations pursue environmental sustainability, particularly by enabling sustainable resource management. This study analyzes multiple case studies to understand how different organizations use IS to address sustainability challenges, providing valuable insights into IS's role in environmental and social sustainability (Hedman and Valentin, 2019).

These works collectively highlight the multifaceted role of IS in fostering sustainability, from supporting green business practices to promoting ethical management of resources. For further exploration, these authors and topics are excellent starting points for deeper research into the intersection of IS success and sustainability management. These references collectively cover various facets of evaluating IS success from a sustainability management perspective, from decision support to CSR, green supply chain management, and ethical considerations.

3. Methodology

Given that the purpose of this research is to develop an enhanced model for information systems evaluation success from the sustainable management perspective, we described our detailed research approach in this section.

Developing a model to evaluate information systems success from a sustainability management perspective involves integrating traditional IS evaluation metrics with sustainability criteria. This requires an understanding of both IS performance and the broader environmental, social, and economic impacts of sustainability. In the following, we present a step-by-step methodology for developing such a model.

The objective is to develop a model to assess the effectiveness and success of IS from a sustainability viewpoint. The scope is to identify which sustainability dimensions (environmental, social, and economic) will be incorporated and how these intersect with IS success factors.

The methodology applied has the following structure. For the goal of developing a new model, the study may be exploratory and for the aim to evaluate existing systems, it may be descriptive and evaluative. The research approach, because we are proposing a conceptual model, it may primarily be qualitative. The review of existing literature on the relationship between sustainability management and information systems discusses frameworks and models used in evaluating IS success (e.g., DeLone & McLean IS Success Model) and identifies gaps related to sustainability dimensions such as environmental, social, and economic factors. For model development, we did the identification of variables for evaluating information system success from a sustainability perspective: metrics for environmental performance, social impacts, economic outcomes, and stakeholder satisfaction. In terms of model structure, we present a conceptual model that incorporates both traditional IS success dimensions (system quality, service quality, user satisfaction) and sustainability-related factors. By following this methodology structure (*see* Figure 1. Methodology structure), we ensure that our material presents a well-organized and rigorous approach to evaluating information systems success through a sustainability management lens.

The existing IS Success Models are well-established IS success frameworks, such as the DeLone and McLean IS Success Model, the Technology Acceptance Model (TAM), and the Information Systems Success Model (ISSM). In order to integrate

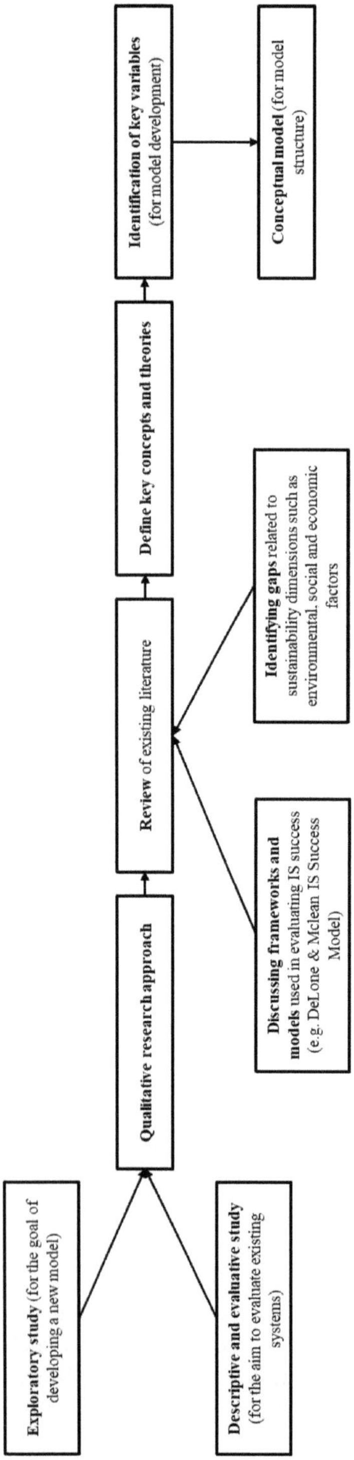

Figure 1. Methodology structure.

Sustainability and IS Success, we wish to synthesize insights from both domains to define the dimensions and metrics for evaluating IS success in a sustainability context. This may include: Environmental Impact - resource consumption, waste reduction, energy efficiency, carbon footprint; Social Impact - stakeholder engagement, community development, user satisfaction, organizational transparency; Economic Impact - cost-effectiveness, return on investment, profitability, long-term financial sustainability.

Evaluating information system success from a sustainability management perspective is a growing area of research. Below are some bibliographical references that discuss models and frameworks for assessing IS success, with a focus on sustainability and long-term organizational and environmental goals. DeLone and McLean foundational paper revisits the DeLone and McLean IS success model, which can be adapted for evaluating IS in the context of sustainability management by considering environmental and social dimensions of IS performance (DeLone and McLean, 2003). Hevner paper introduces design science research in IS, which can be applied to developing IS models that support sustainability management by integrating environmental and social criteria into the success metrics (Hevner et al., 2004). Molla and Licker, while focused on e-commerce, provide insights into evaluating IS success using sustainability-oriented measures, especially in resource-constrained environments (Molla and Licker, 2005). Cheng and Hu focus on the relationship between IS success and sustainability, providing empirical evidence from case studies (Cheng and Hu, 2007). Sahu and Sahu paper examines the intersection of sustainability and IS, proposing frameworks for evaluating the success of IS in contributing to sustainable practices in organizations (Sahu and Sahu, 2015). Tao and Cheng propose a new framework, specifically for evaluating information systems in terms of their sustainability performance, emphasizing both environmental and social factors (Tao and Cheng, 2021). These references provide a solid foundation for understanding how IS success can be evaluated from a sustainability management perspective, and they introduce frameworks that integrate traditional IS success metrics with sustainable development goals.

Based on the literature review, we identified evaluation criteria that span both IS success and sustainability dimensions. These criteria might include: Effectiveness - how well the system supports organizational sustainability goals; Efficiency - resource utilization and operational/efficiency, considering sustainability; User Satisfaction, including social aspects like inclusivity and accessibility; System Quality - reliability, scalability, and adaptability to changing sustainability needs; Impact on Organizational Sustainability Goals - contribution to environmental, social, and economic sustainability; Stakeholder Value - how the system benefits different stakeholders (e.g., employees, customers, society, the environment); Economic Performance - cost savings, efficiency gains, and value creation over time.

4. Results and discussions

As an outcome of the literature review, within the following paragraphs, the main ideas and reasoned thoughts are underlined with the final purpose of developing a new enhanced model for information systems evaluation success.

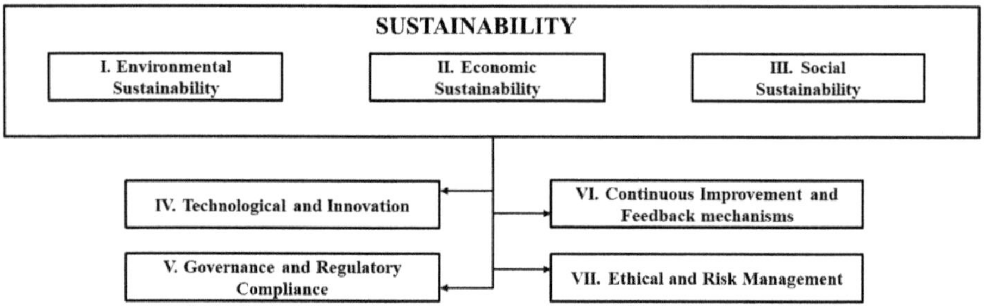

Figure 2. Proposed model.

The model should reflect both the dimensions of IS success and sustainability goals. To develop an enhanced model for evaluating the success of an information system from the sustainability management perspective, it is essential to consider a multidimensional framework that incorporates sustainability's three pillars: environmental, social, and economic (see Figure 2. Proposed model). Below are the dimensions to evaluate an IS's success within the context of sustainability management based on the above-mentioned references: DeLone and McLean, 2003; Hevner et al., 2004; Molla and Licker, 2005; Cheng and Hu, 2007; Mithas et al., 2011; Sahu and Sahu, 2015.

As can be seen within the above figure, the proposed model is comprised of seven dimensions, and its subdimensions are detailed in the below section.

I. **Environmental Sustainability Dimensions** assess how well the information system contributes to reducing environmental impact and fostering sustainable practices. 1. Energy Efficiency sub-dimension to assess the following: if the IS optimizes energy consumption through hardware and software efficiency; if renewable energy sources are being used for data centers or cloud services. 2. Carbon Footprint sub-dimension to assess the following: how much greenhouse gas (GHG) emission is associated with the IS's operation, including server usage, data processing, and energy consumption; if the system helps monitor, reduce, or offset emissions in operations. 3. Waste Reduction sub-dimension to assess the following: if the IS contributes to reducing waste (e.g., paper, electronic waste); if it supports recycling initiatives, such as e-waste management or paperless operations. 4. Resource Optimization sub-dimension to assess the following: how well does the system manage resources (e.g., energy, raw materials) within the organization's supply chain or operations; if it helps organizations reduce resource consumption or facilitate circular economy practices. 5. Environmental Impact Monitoring and Reporting sub-dimension to assess the following: if the IS provides tools for monitoring environmental metrics, such as energy use, water consumption, and waste production; if it assists in generating reports or insights that drive informed sustainability decision-making.

II. **Economic Sustainability Dimensions** assess the financial viability and long-term economic benefits of the information system, ensuring it aligns with sustainable financial practices. 1. Cost Efficiency sub-dimension to assess the following: how cost-effective is the IS in its lifecycle, from initial investment to operational maintenance and upgrades; if the system enables cost savings through automation, resource optimization, or waste reduction. 2. Return on Investment (ROI) sub-dimension to assess the following: what is the long-term financial return of the IS in relation to its sustainability benefits (e.g., cost savings, improved resource management); if the IS contributes to profitability by opening new revenue streams related to sustainability (e.g., sustainable products, services, or market access). 3. Business Model Innovation sub-dimension to assess the following: if the IS supports new business models or revenue opportunities related to sustainability (e.g., sustainable product offerings, green supply chains). 4. Financial Transparency sub-dimension to assess the following: if the IS assists in the transparent reporting of sustainability-related financial metrics, including sustainable investments and carbon-offsetting initiatives; if it helps in assessing the financial impact of sustainability practices and guide resource allocation.

III. **Social Sustainability Dimensions** focuses on the societal impact of the IS, including its influence on employees, customers, and communities. 1. Employee Wellbeing and Work Environment sub dimension to assess the following: how does the IS contribute to creating a better work environment, including promoting employee health, safety, and work-life balance; if it reduces repetitive or hazardous tasks, improving the overall quality of work for employees. 2. Stakeholder Engagement and Collaboration sub dimension to assess the following: how does the IS enable engagement with stakeholders, including employees, customers, suppliers, and the wider community, on sustainability initiatives; if it fosters collaboration across departments, organizations, or sectors to address sustainability challenges. 3. Access to Information and Empowerment sub dimension to assess the following: if the IS provides access to relevant sustainability information for stakeholders (e.g., data on product sustainability, environmental impacts); if it empowers users (employees, customers) to make more sustainable decisions based on real-time data or transparency. 4. Social Equity and Inclusiveness sub-dimension to assess the following: how does the IS contribute to fostering diversity, equity, and inclusiveness in its operations and within the wider community; if the IS supports access to opportunities for under-represented or disadvantaged groups (e.g., fair labour practices, access to education, or health). 5. Community Impact and Social Responsibility sub dimension to assess the following: if the IS helps the organization contribute to local communities, whether through social programs, charitable initiatives, or partnerships.

IV. **Technological and Innovation Dimensions** assesses how the IS integrates emerging technologies to support sustainability goals, as well as its ability to innovate within sustainable contexts. 1. Adaptability to Technological Change sub-dimension to assess the following: how adaptable is the IS to emerging technologies that enhance sustainability (e.g., AI for energy optimization, IoT for resource tracking); if the IS enables the integration of new innovations that improve environmental impact, social inclusion, or operational efficiency.

2. Scalability and Flexibility sub dimension to assess the following: if the IS scales to support growing sustainability goals across multiple regions, sectors, or markets; if the system is flexible enough to accommodate different sustainability strategies, business models, and regulatory requirements. 3. Innovation in Sustainability Solutions sub-dimension to assess the following: if the IS fosters innovation in sustainability, either by providing new solutions (e.g., tools for better resource management) or enabling new business opportunities (e.g., green technology development); if it facilitates the research and development of sustainable products or services that minimize the environmental impact or enhance social wellbeing.

V. **Governance and Regulatory Compliance Dimensions** assesses how well the IS supports compliance with sustainability-related regulations and industry standards. 1. Compliance with Sustainability Regulations sub dimension to assess the following: if the IS helps the organization comply with national or international environmental regulations (e.g., carbon emissions reporting, waste disposal laws); if it assists in adhering to labor laws, ethical sourcing practices, or industry-specific sustainability standards (e.g., ISO 14001, Global Reporting Initiative). 2. Data Privacy and Security in Sustainability Contexts sub dimension to assess the following: how does the IS ensure the security and privacy of sustainability-related data, especially when dealing with sensitive environmental or social data; if the system protects personal data while providing transparency and accountability in sustainability reporting. 3. Sustainability Reporting and Transparency sub-dimension to assess the following: if the IS enables transparent and accurate sustainability reporting, helping the organization meet stakeholder expectations and regulatory requirements; if it integrates sustainability metrics into corporate governance and internal decision-making processes.

VI. **Continuous Improvement and Feedback Mechanisms** assess how the IS supports ongoing sustainability improvement and responsiveness to emerging challenges. 1. Monitoring and Reporting of Sustainability Metrics sub-dimension to assess the following: if the IS provides real-time monitoring of sustainability performance, enabling rapid adjustments to operations; how effectively does it support reporting on progress toward sustainability goals (e.g., carbon neutrality, zero waste). 2. Feedback Loops and Adaptation sub-dimension to assess the following: if the IS facilitates feedback mechanisms that allow for continuous improvement in sustainability practices; if it can track the effectiveness of sustainability initiatives and suggest changes or optimizations based on collected data. 3. Stakeholder Feedback and Engagement sub-dimension to assess the following: how does the IS incorporate stakeholder feedback (e.g., employee surveys, customer insights) to improve sustainability efforts; if the system allows for iterative improvements in processes or products based on stakeholder input.

VII. **Ethical and Risk Management Dimensions** assess how the IS addresses the ethical implications and risks related to sustainability. 1. Ethical Decision-Making and Governance sub-dimension to assess the following: if the IS supports ethical decision-making related to sustainability, such as avoiding greenwashing or ensuring fair labor practices; if it provides tools or frameworks for ethical assessments in sustainability-related projects or investments; 2. Risk Management in the Sustainability sub-dimension to assess the following: how does the IS help identify, assess, and mitigate sustainability-related risks (e.g., climate change risks, supply chain disruptions); if the system assists in managing environmental, social, and economic risks proactively.

The enhanced model for evaluating IS success from a sustainability management perspective focuses on comprehensive dimensions that span environmental, economic, social, technological, governance, and ethical factors. By considering these dimensions, organizations can assess the broader, long-term impacts of their IS in promoting sustainable practices across all levels of operations. The framework ensures that IS contribute not only to business success but also to the achievement of global sustainability goals.

When evaluating the success of information systems from a sustainability perspective, there aren't many well-known, "notorious" or universally recognized surveys specifically tailored to this combination of topics. However, we can adapt existing, prominent IS evaluation models and sustainability frameworks to assess the relationship between IS success and sustainability outcomes. In the following section, we present a few surveys and frameworks that have been used for evaluating IS success, which can be adapted to focus on sustainability.

The DeLone and McLean IS Success Model (1992) is one of the most widely recognized frameworks for evaluating IS success. While this model doesn't specifically focus on sustainability, it provides a comprehensive set of dimensions that can be adapted for a sustainability context. The original dimensions adapted for sustainability are the following: System Quality - how well the system operates in terms of energy consumption, resource efficiency, and scalability to support sustainability goals; Information Quality - how effectively the system provides accurate, relevant data to track sustainability metrics (e.g., carbon emissions, energy use); Use - how often and to what extent the system is utilized for sustainability-related tasks such as resource management, waste tracking, or regulatory compliance; User Satisfaction - how satisfied users are with the system's role in supporting sustainability goals and initiatives; Net Benefits - how well the system contributes to organizational sustainability objectives, including financial (e.g., cost savings from energy efficiency), environmental (e.g., reduced carbon footprint), and social outcomes (e.g., employee engagement or community development) (DeLone and McLean, 1992).

The Information Systems Success Model (ISSM), developed by DeLone and McLean in later years, expands on the original model to incorporate the wider impacts of IS, including factors like user satisfaction and net benefits. It provides a more comprehensive framework for evaluating IS success, making it adaptable to sustainability goals. The original dimensions adapted for sustainability are the following: System Quality - measures of the system's technical efficiency, scalability, and how well it supports resource efficiency and environmental sustainability; Information Quality - the quality of information provided by the system in terms of sustainability data accuracy (e.g., environmental impact reports); Service Quality -

how well the system's support and maintenance contribute to ongoing sustainability efforts (e.g., system updates for new sustainability regulations); Use - the frequency and depth to which the system is used in sustainability-related tasks; User Satisfaction - users' satisfaction with the system in helping to achieve sustainability outcomes; Net Benefits - the overall impact of the system on organizational sustainability goals, including cost savings, improved environmental performance, and enhanced stakeholder engagement (DeLone and McLean, 2003).

The Technology Acceptance Model (TAM) is frequently used to assess user acceptance of new technologies. By adapting TAM to evaluate how well information systems support sustainability goals, we can gauge whether users are more likely to embrace systems that improve sustainability performance.

While there aren't many "notorious" surveys explicitly combining IS success and sustainability, we can adapt existing frameworks like DeLone and McLean's IS Success Model, TAM, GRI Standards, and Sustainable IT Tools to assess how information systems contribute to environmental, social, and economic sustainability outcomes. By tailoring these models to our organization's specific sustainability goals, we can generate insights into how well our IS supports broader sustainability objectives. For survey design considerations we have the following - The *independent variables are system efficiency, system quality, alignment with sustainability goals, stakeholder engagement, and data quality*. The *dependent variables* are *environmental impact, social impact, economic performance, user satisfaction, and long-term sustainability success*. The *moderating variables* are *organizational culture, industry sector, and technology adoption.* The measurement scales are Likert scales (e.g., 1–5 or 1–7 scale) for capturing user satisfaction, system effectiveness, and perceived sustainability impacts. Hereafter, we present several hypotheses we could test in a survey-based study.

Considering the literature review and the main research purpose, the following research hypotheses were drawn:

A. **Hypothesis on Environmental Impact and IS Efficiency:** *H.1 Higher system efficiency (e.g., energy consumption, resource usage) of an information system is positively correlated with reduced environmental impact (e.g., lower carbon footprint, reduced waste generation). H.2 Efficient IS that optimizes resource use or operates on green technologies should result in lower environmental costs* (Davis, 1989; DeLone and McLean,1992; DeLone and McLean, 2003; Sustainable IT Consortium, 2019; Global Reporting Initiative (GRI), 2021).

B. **Hypothesis on IS Effectiveness and Sustainability Goals:** *H.3 The effectiveness of an information system in supporting organizational sustainability goals is positively correlated with the overall success of sustainability initiatives within the organization. H.4 A system's ability to track, monitor, and optimize sustainability metrics is expected to directly influence the achievement of sustainability targets* (Davis, 1989; DeLone and McLean, 1992; DeLone and McLean, 2003; Sustainable IT Consortium, 2019; Global Reporting Initiative (GRI), 2021).

C. **Hypothesis on Social Impact and User Satisfaction:** *H.5 User satisfaction with an information system is positively correlated with its contribution to social sustainability outcomes, such as employee well-being, stakeholder engagement, and community development. H.6 IS that prioritize social aspects, such as inclusivity and fairness, are likely to result in higher user satisfaction and positive social impact targets* (Davis, 1989; DeLone and McLean, 1992; DeLone and McLean, 2003; Sustainable IT Consortium, 2019; Global Reporting Initiative (GRI), 2021).
D. **Hypothesis on System Quality and Sustainability Outcomes:** *H.7 Higher quality information systems (e.g., reliability, scalability, and flexibility) contribute to better sustainability performance (environmental, social, and economic outcomes). H.8 Robust systems that are reliable and adaptable help organizations optimize processes that contribute to sustainability across different dimensions targets* (Davis, 1989; DeLone and McLean, 1992; DeLone and McLean, 2003; Sustainable IT Consortium, 2019; Global Reporting Initiative (GRI), 2021).
E. **Hypothesis on Economic Performance and IS Cost-Effectiveness:** *H.9 The economic performance of an information system (e.g., cost savings, ROI) is positively correlated with its sustainability performance, as it drives cost-effective resource management and reduces waste. H.10 Economically efficient systems are often more sustainable by reducing unnecessary costs and optimizing resource use, leading to a direct impact on both financial and sustainability goals targets* (Davis, 1989; DeLone and McLean, 1992; DeLone and McLean, 2003; Sustainable IT Consortium, 2019; Global Reporting Initiative (GRI), 2021).
F. **Hypothesis on Stakeholder Value and IS Success:** *H.11 Information systems that are perceived to create value for a wider range of stakeholders (e.g., employees, customers, suppliers, local communities) contribute more significantly to the long-term success of sustainability management. H.12 Systems that benefit a wide array of stakeholders in sustainable ways will be more likely to foster trust, satisfaction, and organizational success, which are crucial for long-term sustainability targets* (Davis, 1989; DeLone and McLean, 1992; DeLone and McLean, 2003; Sustainable IT Consortium, 2019; Global Reporting Initiative (GRI), 2021).
G. **Hypothesis on Organizational Alignment with Sustainability Strategy:** *H.13 Information systems that are aligned with an organization's sustainability strategy are more likely to be perceived as successful in achieving both IS and sustainability outcomes. H.14 A strong alignment between IS and sustainability goals ensures that the technology supports the organization's strategic objectives, leading to improved overall performance targets* (Davis, 1989; DeLone and McLean, 1992; DeLone and McLean, 2003; Sustainable IT Consortium, 2019; Global Reporting Initiative (GRI), 2021).
H. **Hypothesis on Adaptability and Long-Term Sustainability:** *H.15 Information systems that are adaptable to changing sustainability regulations*

and evolving environmental and social needs are more likely to contribute to long-term organizational sustainability. H.16 Systems that can evolve with new sustainability trends and regulatory changes ensure that an organization can maintain compliance and continue to meet sustainability objectives over targets (Davis, 1989; DeLone and McLean, 1992; DeLone and McLean, 2003; Sustainable IT Consortium, 2019; Global Reporting Initiative (GRI), 2021).

I. **Hypothesis on IS Integration with Sustainability Reporting:** *H.17 The integration of information systems with sustainability reporting tools positively influences an organization's ability to track, report, and improve sustainability performance. H.18 IS that facilitate data collection and analysis for sustainability reports enhance an organization's transparency, accountability, and continuous improvement in sustainability practices* (Davis, 1989; DeLone and McLean, 1992; DeLone and McLean, 2003; Sustainable IT Consortium, 2019; Global Reporting Initiative (GRI), 2021).

J. **Hypothesis on Data Quality and Decision-Making for Sustainability:** *H.19 High-quality, accurate, and timely data provided by an information system enhances decision-making capabilities related to sustainability practices and initiatives. H.20 Access to reliable data is crucial for making informed decisions regarding sustainability, helping organizations optimize their sustainability strategies and actions targets* (Davis, 1989; DeLone and McLean, 1992; DeLone and McLean, 2003; Sustainable IT Consortium, 2019; Global Reporting Initiative (GRI), 2021).

By testing these hypotheses through a well-designed survey, organizations can gain insights into how their information systems contribute to sustainability efforts across environmental, social, and economic dimensions.

When designing survey to evaluate the success of information systems from a sustainability management perspective, it is crucial to consider both traditional IS success factors (e.g., system quality, user satisfaction) and sustainability-specific metrics (e.g., environmental, social, and economic impacts). The items refer to the following dimensions: System Efficiency and Environmental Impact - to assess how well the system contributes to environmental sustainability by reducing resource consumption, energy use, and waste; System Quality - to focus on the quality of the information system, including its reliability, usability, and performance, which indirectly affect sustainability outcomes; User Satisfaction and Social Impact - to evaluate the social aspects of the system, such as user satisfaction, inclusivity, and its contribution to social sustainability (e.g., employee well-being, community engagement); Economic Performance and Cost Efficiency - to assess how the IS contributes to the financial sustainability of the organization, including cost savings, profitability, and overall economic benefits.

Evaluating the success of an information system from a sustainability management perspective requires considering both traditional success metrics (like efficiency and effectiveness) and environmental, social, and economic sustainability dimensions.

Below, is the final proposed framework to enhance model evaluation from this perspective based on the above mentioned references: DeLone and McLean, 2003; Hevner et al., 2004; Molla and Licker, 2005; Cheng and Hu, 2007; Mithas et al., 2011; Sahu and Sahu, 2015.

Apart from adding the usual main dimensions of sustainability, the newly proposed enhanced model integrates the key performance indicators (KPIs) aligned with sustainability goals and evaluates how well the information system contributes to sustainable practices. The model dimensions and sub-dimensions are outlined and described as below (Figure 3):

I. **Sustainability Dimensions Integration** should incorporate the three main pillars of sustainability: environmental, economic, and social aspects. This helps in assessing how well the information system supports the long-term sustainability of the organization and its broader impact on society and the environment. 1. Environmental Sustainability with the following sub-dimensions: (a) Energy Consumption to evaluate how efficiently the IS uses energy (e.g., server optimization, cloud computing energy use). (b) Carbon Footprint to assess the emissions directly or indirectly associated with the operation of the IS. (c) Waste Reduction to evaluate the extent to which the IS helps in reducing waste (e.g., paper, electronic waste management). (d) Resource Efficiency to measure how the system contributes to resource management and conservation (e.g., water, energy, raw materials). 2. Economic Sustainability with the following sub-dimensions: (a) Cost Efficiency to evaluate the total cost of ownership (TCO), including initial investment, maintenance, and operational costs. (b) Return on Investment to assess long-term financial gains from the system (e.g., cost savings, operational efficiency, competitive advantage). (c) Revenue Generation to identify how the IS may contribute to new business models, services, or revenue streams, particularly in green and sustainable sectors. (d) Market Adaptability to evaluate the system's ability to adapt to market changes and regulatory demands (e.g., compliance with environmental standards). 3. Social Sustainability with the following sub-dimensions: (a) Stakeholder Engagement to measure how well the system engages various stakeholders (employees, customers, suppliers) in sustainability efforts. (b) Employee Wellbeing to evaluate how the system supports social factors like work-life balance, health, and safety through automation or data management. (c) Community Impact to assess how the system benefits local communities, including employment opportunities, social responsibility projects, and local development.

II. **Success Metrics Based on IS Performance** can assess how effectively an information system meets both organizational and sustainability goals. 1. Information System Effectiveness with the following sub-dimensions: (a) Decision Support to evaluate how well the IS provides relevant data for sustainability-related decisions (e.g., energy consumption analysis, sustainability reporting). (b) Efficiency Gains to measure improvements in operational efficiency through automation, resource management, or waste reduction. (c) User Satisfaction to evaluate user experience and satisfaction levels, especially

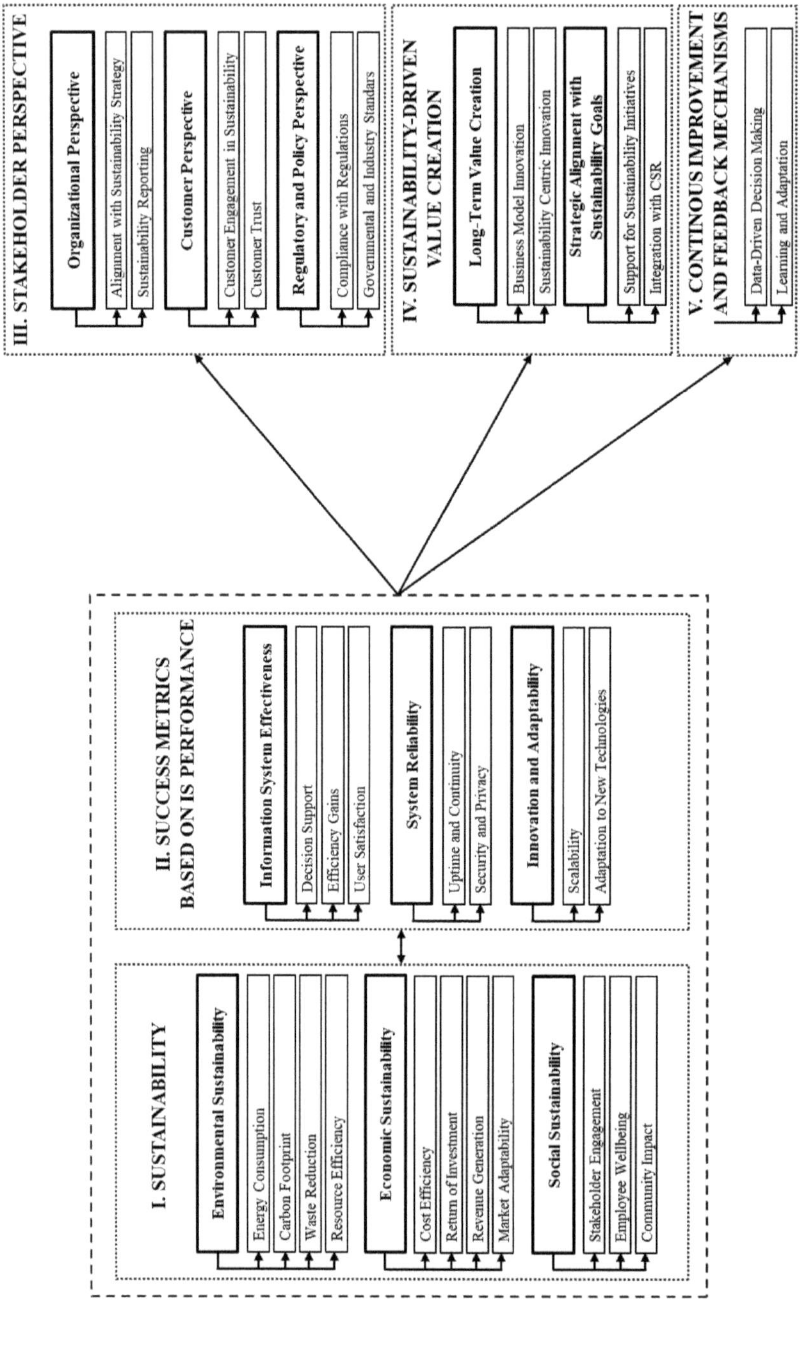

Figure 3. Proposed enhanced model.

among those managing sustainability-related data or stakeholders. 2. System Reliability with the following sub-dimensions: (a) Uptime and Continuity to evaluate system reliability, focusing on the importance of availability for sustainability operations. (b) Security and Privacy to assess how secure the system is in handling sensitive environmental or sustainability data and the implications of data breaches. 3. Innovation and Adaptability with the following sub-dimensions: (a) Scalability to evaluate how easily the IS can scale to accommodate growing sustainability goals or expanding operations. (b) Adaptation to New Technologies to assess the system's ability to integrate with emerging technologies (e.g., IoT for sustainability monitoring, AI for environmental impact predictions).

III. **Stakeholder Perspective.** Incorporating various stakeholder perspectives helps in evaluating how the IS serves different sustainability-related goals. 1. Organizational Perspective with the following sub-dimensions: (a) Alignment with Sustainability Strategy to evaluate how well the IS supports the organization's broader sustainability strategy (e.g., zero-emissions goals, circular economy principles). (b) Sustainability Reporting to evaluate the IS's role in producing transparent, accurate, and actionable sustainability reports. 2. Customer Perspective with the following sub-dimensions: (a) Customer Engagement in Sustainability to assess the system's role in enabling customers to engage with sustainability initiatives (e.g., carbon footprint tracking tools). (b) Customer Trust to evaluate how the IS fosters trust through transparency, such as providing access to real-time sustainability data. 3. Regulatory and Policy Perspective with the following sub-dimensions: (a) Compliance with Regulations to evaluate how well the IS helps ensure compliance with local and international sustainability regulations (e.g., emissions reporting, sustainability disclosures). (b) Governmental and Industry Standards to measure how the system aligns with recognized sustainability standards (e.g., ISO 14001 for environmental management systems).

IV. **Sustainability-Driven Value Creation** assess how the IS contributes to sustainability-driven value creation. 1. Long-Term Value Creation with the following sub-dimensions: (a) Business Model Innovation to assess the extent to which the IS supports new business models that prioritize sustainability (e.g., offering green products, sustainable supply chains). (b) Sustainability-centric innovation to evaluate how the IS fosters innovations that contribute to sustainable development (e.g., new products, services, or technologies that address sustainability issues). 2. Strategic Alignment with Sustainability Goals with the following sub-dimensions: (a) Support for Sustainability Initiatives to measure the extent to which the IS enables the achievement of specific sustainability objectives (e.g., reducing environmental impact, improving social equity). (b) Integration with Corporate Social Responsibility to evaluate how the IS contributes to the company's CSR activities and initiatives.

V. **Continuous Improvement and Feedback Mechanisms** to evaluate how the IS supports continuous sustainability improvement. 1. Data-driven decision-making to evaluate how well the system collects, analyzes, and presents data that supports continuous sustainability improvements and to assess how feedback loops within the system help refine sustainability efforts over time (e.g., through monitoring and adjusting sustainability practices). 2. Learning and Adaptation to measure the IS's ability to learn from sustainability outcomes and incorporate lessons into system upgrades or processes and to evaluate the level of adaptability within the system to meet emerging sustainability challenges or incorporate new best practices.

This framework offers a comprehensive, multi-dimensional approach to evaluating information system success from a sustainability management perspective. By considering environmental, social, and economic sustainability metrics alongside traditional IS performance indicators, organizations can ensure that their information systems contribute to long-term, sustainable development goals.

5. Conclusions

The necessity of a model for evaluating the success of information systems from the sustainability management perspective is growing increasingly important due to the integration of environmental, social, and economic concerns in modern business operations. Traditional models of IS success typically focus on operational performance metrics like system usage, user satisfaction, and financial outcomes. However, they often overlook the broader implications of how IS contribute to sustainability goals, which is a critical issue in today's environmentally-conscious and socially responsible business environment.

Evaluating information systems from a sustainability perspective emphasizes long-term value creation rather than just immediate profits. It can uncover ways in which IS can enable resource efficiency, enhance social responsibility, or support circular economy initiatives, thereby fostering long-term organizational sustainability.

Many organizations have embraced corporate sustainability, but IT strategies and systems may not always be aligned with these initiatives. A sustainability-oriented IS evaluation model can bridge this gap by providing clear metrics and frameworks for integrating sustainability considerations into IT decision-making processes.

IS have non-financial impacts on sustainability, such as promoting social inclusion, improving governance structures, and supporting environmental initiatives. A comprehensive evaluation model can account for these dimensions, ensuring that IS success is measured in terms of all three pillars of sustainability: environmental, social, and economic.

In conclusion, as sustainability becomes a central concern for businesses, an evaluation model that incorporates sustainability management into IS success is essential. Such a model would not only help businesses improve operational performance but also support their long-term environmental and social responsibility, driving a more comprehensive and forward-thinking approach to information systems success.

References

Bélanger, F. and Crossler, R. E. (2011). Privacy in the digital age: a review of the literature. Journal of the Association for Information Systems 12(5). https://doi.org/10.2307/41409971.

Bharati, P. and Chaudhury, A. (2015). Sustainability and information systems: a literature review and directions for future research. Journal of Strategic Information Systems 24(2). https://doi.org/10.3390/su12083337.

Cheng, M. Y. and Hu, H. C. (2007). An empirical study of information systems success on sustainability. Journal of Global Information Management 15(2). https://doi.org/10.3390/su12083337.

Davis, F. D. (1989). Perceived usefulness, perceived ease of use, and user acceptance of information technology. MIS Quarterly. https://doi.org/10.2307/249008.

DeLone, J. E. and M. M. B. (2012). The sustainable business value of information systems. Journal of Strategic Information Systems 21(1). https://doi.org/10.1080/07421222.2003.11045748.

DeLone, W. H. and McLean, E. R. (1992). Information systems success: The quest for the dependent variable. Information Systems Research 3(1). https://doi.org/10.1287/isre.3.1.60.

DeLone, W. H. and McLean, E. R. (2003). The DeLone and McLean model of information systems success: a ten-year update. Journal of Management Information Systems 19(4). https://doi.org/10.1080/07421222.2003.11045748.

Fitzgerald, G. and Russo, N. L. (2012). Managing information systems for sustainability: strategic approaches and implementation challenges. Information & Management 49(6). https://doi.org/10.1111/j.1365-2575.2012.00410.x.

Global Reporting Initiative (GRI). (2021). GRI Standards. GRI Website. https://www.globalreporting.org.

Hedman, J. and Valentin, E. K. (2019). Information systems and sustainability: a multiple case study. Information Systems Journal 29(5). https://doi.org/10.1111/isj.12489.

Hevner, A. R., March, S. T., Park, J. and Ram, S. (2004). Design science in information systems research. MIS Quarterly 28(1). https://dl.acm.org/doi/10.5555/2017212.2017217.

Hevner, A. R. and Chatterjee, S. (2010). Design Research in Information Systems: Theory and Practice. Springer.

ISO 14001:2015 Environmental Management Systems (EMS) Standard.

Melville, N. P. (2010). Information systems innovation for environmental sustainability. MIS Quarterly 34(1). https://doi.org/10.4236/jep.2019.102014.

Melville, N. P. (2010). Information systems innovation for environmental sustainability. Journal of the Association for Information Systems 11(11). https://doi.org/10.2307/20721412.

Mithas, S., Ramasubbu, N. and Sambamurthy, V. (2011). How information management capabilities influence firm performance. MIS Quarterly 35(1). https://doi.org/10.2307/23043496.

Molla, A. and Licker, P. S. (2005). E-commerce adoption in developing countries: a model and instrument. Information & Management 42(6). https://doi.org/10.1016/j.im.2004.09.002.

Ng, E. K. (2020). Sustainable Information Systems: Frameworks for Assessing Their Impact on Corporate Social Responsibility. Master's Thesis, University of Melbourne.

Sahu, S. and Sahu, S. (2015). Towards sustainable information systems: a critical review and future directions. Sustainability 7(5): 6247–6274.

Sustainable IT Consortium. (2019). Sustainable IT Assessment Framework.

Tao, Y. and Cheng, Z. (2021). Sustainable information systems: a comprehensive framework for evaluation. Journal of Information Science 47(5): 631–648.

United Nations Global Compact & Accenture. (2019). Sustainability and the Digital Age: The Role of Information Systems in Achieving the SDGs.

World Economic Forum. (2020). Digital Transformation and Sustainability: Measuring the Success of Information Systems for a Sustainable Future.

Chapter 6

Empowering Workforce Development

Green Intelligence through AI and STEM Skills

*Larisa Ivascu,** *Dana Fatol* and *Nicoleta Mirea*

1. Introduction

In the face of rapid technological advancements and intensifying environmental challenges, workforce sustainability has become a strategic priority for organisations and societies worldwide. At the core of this effort lies the concept of Green Intelligence, which encompasses the knowledge, skills, and competencies essential to drive environmentally sustainable practices. By integrating environmental awareness, technical expertise, and adaptive capabilities, Green Intelligence empowers organisations to navigate the complexities of transitioning toward a green economy. As sustainability goals are taken to the forefront, addressing the growing skills gap is critical to foster a workforce capable of meeting these demands (Raman et al., 2024; Yin et al., 2023).

Integral to achieving this goal are STEM skills - Science, Technology, Engineering, and Mathematics - which serve as foundational pillars for innovation and problem solving. STEM competencies enable professionals to design and implement scalable sustainable solutions, bridging the gap between technological

Politehnica University of Timisoara, Faculty of Management in Production and Transportation, 14 Remus Street, Timisoara, Romania.
Emails: dana.fatol@student.upt.ro, nicoleta.mirea@student.upt.ro
* Corresponding author: larisa.ivascu@upt.ro

DOI: 10.1201/9781003518990-6

advancements and environmental imperatives. Research underscores the need for an interdisciplinary approach to STEM education, integrating sustainability concepts to prepare people for addressing complex global challenges effectively.

The integration of Artificial Intelligence (AI) with STEM and Green Intelligence presents transformative opportunities for workforce development. AI-powered systems facilitate personalised learning, adaptive assessments, and real-time feedback, accelerating the acquisition of critical competencies. Furthermore, AI-driven frameworks for incorporating STEM into Green Intelligence development offer a pathway to cultivate a new generation of innovators poised to address urgent sustainability challenges. By aligning technological capabilities with sustainability-driven education, these frameworks position organisations to meet evolving environmental and economic demands.

Despite these advances, talent management systems today face significant challenges. Organisations operate within dynamic markets shaped by globalisation, technological disruption, and environmental imperatives. Talent management has become a strategic necessity, requiring innovative approaches to recruitment, development, and retention to maintain competitive advantages (Fatol et al., 2020). Furthermore, the persistent gap between academia and business adds complexity to implementing sustainable workforce strategies. While academia emphasises theoretical rigor, businesses prioritise practical outcomes, often leading to misalignment. Bridging this divide requires fostering stronger collaborations to align research with real-world applications (Fatol and Robescu, 2023).

This paper examines the intersection of Green Intelligence, STEM skills, and AI in shaping sustainable workforce development. It synthesises current research, explores practical applications, and proposes strategies for integrating these elements into talent management systems. By leveraging these interdisciplinary approaches, organisations can enhance resilience, drive innovation, and contribute to global sustainability, positioning themselves for success in an increasingly complex and dynamic world.

2. The role of Green Intelligence in sustainable workforce strategies

2.1 Understanding Green Intelligence in workforce sustainability

In the context of workforce sustainability, Green Intelligence is an emerging concept that captures the knowledge, skills, and competencies required to navigate and drive the transition to environmentally sustainable practices within organisations. This multidimensional approach integrates environmental awareness, technical expertise, and adaptive capabilities to address the dynamic needs of a green economy.

Based on recent research, we can define Green Intelligence in workforce sustainability as the collective capacity of an organisation's workforce to understand, implement, and innovate sustainable practices, including both specialised green skills and the ability to apply existing competencies to environmentally focused roles and assignments. This type of intelligence enables workers to contribute effectively to an organisation's environmental goals while adapting to the rapidly changing landscape of green jobs and technologies. The concept of Green Intelligence has gained significant progress in recent years, driven by the urgent need to address climate change and the shift towards sustainable business practices. According to the 2024 LinkedIn Global Green Skills Report, the demand for green talent rose by 11.6% between 2023 and 2024, while the supply only increased by 5.6%, highlighting a critical skills gap in the workforce (Beamery, 2024a; LinkedIn Talent Solutions, 2024).

Green Intelligence includes a broad spectrum of skills, ranging from technical expertise in renewable energy and sustainable construction to more general competencies such as critical thinking and complex problem solving. Research indicates that process skills and cross-functional skills are particularly crucial in enabling workers to adapt to the innovation-driven nature of the green transition (OECD, 2024).

The development of Green Intelligence in the workforce has significant implications for organisational sustainability and competitiveness. Companies that prioritise the development of green skills among their employees are better positioned to drive innovation in sustainable products and services and attract environmentally conscious consumers and investors.

To address the green skills gap, organisations can implement various strategies. These include the following:

- *Upskilling and reskilling initiatives*: Companies are investing in targeted training programmes to develop green competencies among existing employees (Beamery, 2024a).
- *Skills-based hiring models*: Organisations are adopting more flexible hiring practices that focus on transferable skills and potential rather than specific green credentials (Beamery, 2024a).
- *Collaboration with educational institutions*: Creating partnerships between industry and academia to align curriculum development with the increasing needs of the green economy (Simmonds and Lally, 2024).
- *Policy interventions*: Governments are implementing policies to incentivise the development of green skills and create supportive frameworks for the transition to a sustainable workforce (Cheng et al., 2023).

The concept of Green Intelligence extends beyond the "green jobs" to incorporate the broader workforce. As found in recent research, green skills will soon become "business-as-usual", integrated into various roles across different sectors, much like digital skills have become widespread (Deloitte UK, 2024).

2.2 STEM skills as pillars of sustainable workforce practices

STEM skills refer to the integration of Science, Technology, Engineering, and Mathematics competencies into talent management strategies that promote long-term environmental, social, and economic sustainability. These skills are crucial for developing innovative solutions to global challenges and fostering a workforce capable of adapting to rapidly evolving technological and environmental landscapes (Chiangpradit, 2024; Valdes-Ramirez et al., 2024).

Recent research has highlighted the growing importance of STEM skills in sustainable workforce development. A 2024 study by Hakansson Lindqvist emphasises the critical role of STEM education in cultivating a holistic understanding of the interconnections between innovation and sustainability dimensions (Valdes-Ramirez et al., 2024). This integration is essential for preparing students to contribute effectively to solving environmental, economic, and social sustainability problems.

The concept of STEM skills as pillars of sustainable workforce practices involves several key components:

- *Interdisciplinary approach*: STEM skills foster an integrated understanding of complex sustainability issues, allowing professionals to approach problems from multiple perspectives (Valdes-Ramirez et al., 2024).
- *Innovation and problem-solving*: STEM education equips people with the ability to develop and implement sustainable technologies and practices (Chiangpradit, 2024; Scientix, 2024).
- *Data-driven decision-making*: Proficiency in STEM fields enables the use of advanced analytics and data science techniques to support sustainability strategies (Valdes-Ramirez et al., 2024).
- *Adaptability and lifelong learning*: STEM skills provide a foundation for continuous learning and adaptation to emerging technologies and sustainability challenges (Talent Management Institute, 2024).

The implications of integrating STEM skills into sustainable workforce practices are significant. Organisations that prioritise these competencies are better positioned to address environmental concerns, drive innovation, and maintain long-term economic viability (Talent Management Institute, 2024). Furthermore, the demand for green jobs is growing nearly twice as fast as the number of workers with the requisite skills, highlighting the urgent need for STEM education and training in sustainability (World Economic Forum, 2024).

Practical applications of STEM skills in sustainable workforce practices include:

- Developing renewable energy technologies and improving energy efficiency (Becker, 2024; Scientix, 2024).
- Design and implementing sustainable agricultural practices to ensure food security (Scientix, 2024).
- Creating innovative solutions for waste reduction and circular economy initiatives (Chiangpradit, 2024).
- Implementing data-driven approaches to measure and improve sustainability performance (Valdes-Ramirez et al., 2024).
- Recent research also underscores the importance of early STEM education to prepare the future workforce for sustainability challenges. A 2024 study emphasises the need to integrate sustainability concepts into STEM curricula

from an early age, fostering a generation of professionals equipped to address global environmental issues(Chiangpradit, 2024).

2.3 Fostering talent development with Green Intelligence

In the rapidly evolving landscape of human resource management, the integration of environmental considerations into talent development strategies has emerged as a critical focus.

Talent is usually seen as the combination of an individual's skills, knowledge, and potential that can be leveraged to drive organisational success. In the context of Green Intelligence, talent includes not only traditional competencies but also environmental awareness and the capacity to contribute to sustainable business practices (Beamery, 2024b).

Green Talent Management (GTM) has emerged as a key concept in environmentally conscious workforce. GTM refers to the systematic process of identifying, attracting, developing, and retaining employees with the skills and motivation to support an organisation's environmental objectives (Iberdrola, 2024). This approach aligns human resource practices with sustainability goals, creating a synergy between talent development and ecological responsibility.

There are several key components of Green Intelligence in talent development:

- *Environmental competence*: The latest findings indicate that developing environmental competence is crucial for fostering green talent. This involves equipping employees with the knowledge and skills necessary to understand and address environmental challenges within their roles (Mercer, 2024).
- *Sustainable leadership*: Recent studies emphasise the importance of cultivating sustainable leadership qualities. This includes the ability to make decisions that balance economic, social and environmental considerations, driving long-term organisational sustainability (WorkDove, 2024).
- *Innovation for sustainability*: Green Intelligence in talent development also focusses on nurturing innovative thinking orientated towards sustainability. This component encourages employees to develop creative solutions that address environmental concerns while maintaining business efficiency (Rocki, 2024).

The integration of Green Intelligence into talent development strategies has significant implications for workforce sustainability.

- *Enhanced employee engagement*: Recent insights suggest that organizations implementing GTM practices experience higher levels of employee engagement and commitment (Odugbesan et al., 2022).
- *Competitive advantage*: Developing a workforce with strong Green Intelligence helps organisations navigate increasingly demanding environmental regulations and meet evolving consumer expectations (Yu et al., 2024).
- *Improved organisational performance*: Studies indicate a positive correlation between GTM practices and overall organizational performance, including financial metrics.

The concept of fostering talent development with Green Intelligence represents a paradigm shift in human resource management, aligning workforce capabilities with global sustainability imperatives. As organisations continue to grapple with environmental challenges, the integration of Green Intelligence into talent development strategies will likely become increasingly crucial.

3. STEM-focused and AI-enhanced workforce development programmes

3.1 AI-driven integration of STEM in corporate training

The integration of AI into STEM education has become a transformative element in corporate training, introducing innovative methods for skill development and knowledge acquisition.

The interest in the strategic use of AI-powered tools and methodologies to improve STEM-related skills and knowledge within organisational learning environments has gained more ground in recent years. This approach leverages AI's capabilities to create personalised, adaptive, and engaging learning experiences that address the complex and rapidly evolving needs of the modern workforce (Glick et al., 2024; Khawaji, 2024).

The conceptual model for AI-driven STEM integration in corporate trainings covers several key components:

- AI-powered virtual human training modules
- Adaptive learning systems
- Intelligent tutoring systems
- Data analytics for performance assessment
- Augmented and virtual reality simulations

These elements work in harmony to establish a holistic learning ecosystem, fostering the development of essential STEM competencies within the corporate environment (Duterte, 2024; Glick et al., 2024).

Fresh insights have highlighted several key concepts and practical implications of AI-driven STEM integration in corporate training.

- *Enhanced engagement and understanding*: AI-powered tools, such as ChatGPT, have been shown to significantly improve employee engagement and comprehension of complex STEM concepts (Duterte, 2024). This enhanced understanding is crucial for organisations seeking to build a technologically proficient workforce.
- *Personalised learning pathways*: AI algorithms can analyse individual learning patterns and knowledge levels to create customised training content, optimising the learning process for each employee (Glick et al., 2024; Khawaji, 2024).
- *Scalable skill development*: AI-powered virtual human training modules offer a scalable solution for developing essential skills, such as self-regulated

learning strategies, which are critical for success in STEM-related roles (Glick et al., 2024).

- *Real-time feedback and assessment*: Intelligent tutoring systems provide immediate feedback and adaptive assessments, allowing for continuous improvement and more accurate evaluation of employee competencies (Olatunbosun Bartholomew Joseph and Nwankwo Charles Uzondu, 2024).
- *Bridging theory and practice*: Augmented and virtual reality simulations, powered by Artificial Intelligence (AI), enable employees to apply theoretical knowledge in realistic and risk-free environments, bridging the gap between classroom learning and practical application (Khawaji, 2024).

Integration of AI-driven STEM training has significant implications for workforce sustainability. The four most recognised are:

- *Continuous skill adaptation*: AI-powered training systems can rapidly update content to reflect the latest technological advances, ensuring that employees' skills remain relevant in a fast-paced industry (Falebita and Kok, 2024; Olatunbosun, Bartholomew Joseph and Nwankwo, Charles Uzondu, 2024).
- *Improved retention and performance*: Personalised and engaging learning experiences facilitated by AI can lead to better knowledge retention and improved job performance, contributing to long-term employee satisfaction and productivity (Duterte, 2024; Falebita and Kok, 2024).
- *Inclusive learning environments*: AI-driven training can accommodate diverse learning styles and backgrounds, promoting inclusivity and equal opportunities for skill development across the organisation (Duterte, 2024; Falebita and Kok, 2024).
- *Data-driven decision making*: The wealth of data generated by AI-powered training systems can inform strategic workforce planning and identify areas for organisational improvement (Khawaji, 2024; Olatunbosun, Bartholomew Joseph and Nwankwo, Charles Uzondu, 2024).

The AI-driven integration of STEM in corporate training represents a paradigm shift in organisational learning and development. By leveraging AI technologies, companies can create more effective, engaging, and scalable training programmes that address the complex skill requirements of the modern STEM workforce.

3.2 An AI-Driven framework for STEM integration in Green Intelligence development

The convergence of AI and sustainability has emerged as a critical area of research, reflecting the growing need to integrate sustainable practices into AI initiatives (Raman et al., 2024). An AI-driven framework can leverage technologies to enhance the integration of STEM disciplines in developing environmentally sustainable and intelligent systems. Recent studies show that this framework facilitates innovation, promotes energy efficiency, and address global sustainability challenges through the integrated application of AI and STEM education (Portillo-Blanco et al., 2024; Telfort, 2024).

AI has a greater potential in driving sustainable development and green innovation. UNESCO (2024a) emphasise the importance of integrating AI competencies into education to prepare students for the challenges of a rapidly evolving technological landscape (UNESCO, 2024b). This integration is crucial for developing a workforce capable of addressing complex environmental issues through innovative STEM applications.

Our analysis shows that an effective framework should include the following key components:

1. *AI-Enhanced STEM education*: Incorporating AI technologies into STEM curricula to provide personalised learning experiences and foster critical thinking skills necessary for the development of Green Intelligence (Telfort, 2024).
2. *Green AI research*: Focusing on developing environmentally sustainable AI technologies that optimise energy consumption and minimize carbon footprints (Raman et al., 2024).
3. *Interdisciplinary collaboration*: encouraging cross-disciplinary research and development that combines expertise from various STEM fields to address sustainability challenges (Portillo-Blanco et al., 2024).
4. *Ethical considerations*: Integrating ethical principles into the development and application of AI in Green Intelligence to ensure responsible and sustainable innovation (UNESCO, 2024a).

The implications of this framework for the sustainability of the workforce are significant. As the global AI in the education market continues to grow rapidly, there is an increasing demand for professionals who are skilled in both AI and sustainability practices. The framework promotes the development of a workforce that is not only technologically competent, but also environmentally conscious and capable of driving green innovation.

Potential applications of this framework include the following:

- Developing AI-powered systems for the optimization and management (Yin et al., 2023).
- Creating intelligent environmental monitoring solutions using AI and STEM principles (Alamäki et al., 2024).
- Design sustainable AI algorithms and hardware that minimise energy consumption (Raman et al., 2024).

An AI-driven framework for STEM integration in the development of Green Intelligence represents a promising approach to addressing global sustainability challenges. By combining the power of AI with comprehensive STEM education, this framework has the potential to foster a new generation of innovators capable of developing sustainable solutions for a greener future.

4. Green Intelligence metrics in workforce development through STEM and AI

4.1 STEM and AI performance indicators and technologies in Green Intelligence initiatives

In the context of accelerating climate crises and intensifying sustainability efforts, organisations in various industries are focusing on implementing Green Intelligence initiatives (Alzoubi and Mishra, 2024; Jeevaraj et al., 2023). These initiatives, which aim to reduce environmental impact and increase energy efficiency, are supported by well-defined performance indicators and the advanced use of STEM and AI technologies. The role of these technologies is essential to monitor and report environmental performance, optimise processes, and develop strategies to reduce carbon emissions (Belkhir and Elmeligi, 2018; Jeevaraj et al., 2023; Khan et al., 2022; Li et al., 2024).

AI and STEM contribute to a systemic transformation of organisations by integrating intelligent monitoring, analysis, and forecasting solutions (Chen et al., 2023; Franki et al., 2023; Jatana et al., 2024; Mazhar et al., 2023; Rashid and Kausik, 2024; Sarfraz et al., 2024). For example, by using IoT sensors and machine learning algorithms, organisations can assess energy efficiency in real time, anticipate consumption needs, and implement proactive strategies to save resources. Advanced big data-driven analytics platforms also provide decision-making support to reduce environmental risks and accelerate the transition to a circular economy (Kumar et al., 2024; Rashid and Kausik, 2024).

Green Intelligence initiatives are facilitated by a set of performance indicators that measure factors such as energy consumption, CO_2 emissions, resource recycling, and renewable energy use (Cheng et al., 2023; Hasan et al., 2024; Shukla et al., 2024; Tornjanski et al., 2024). These metrics allow organisations to set clear goals, monitor progress, and adapt sustainability strategies on an ongoing basis. Through synergy between STEM technologies, AI, and performance indicators, organisations respond to current legislative and social requirements, but also position themselves as leaders in global efforts to protect the environment.

This integration represents an example of a paradigmatic shift, where emerging technologies and traditional performance management methods converge to generate innovative and sustainable solutions (Bibri et al., 2023; Dzhunushalieva and Teuber, 2024; Ng, 2019; Savon et al., 2020).

In an increasingly complex economic and technological environment, it is essential to identify the specific skills requirements of AI to build a workforce capable of addressing sustainability challenges.[1] Using clear classifications, such as those developed by Lightcast (Borgonovi et al., 2023), classifications of AI skills into "*general*" and "*specific*", it is possible to analyse labour market demand and shape effective education strategies. In this section, this approach aims to understand how AI competencies are linked to STEM education and contribute to the development of

[1] https://t4.oecd.org/skills/oecd-skills-outlook-e11c1c2d-en.htm.

Green Intelligence metrics needed to measure progress in sustainable and innovative development.

4.2 The role of performance indicators in Green Intelligence

Performance indicators are fundamental for assessing the impact of Green Intelligence and STEM and AI initiatives (Table 1). These indicators provide the necessary data for strategic decision making and continuous process adaptation, allowing accurate tracking of organisational progress. In particular, indicators such as energy and digital efficiency and the adoption of green technologies offer insight into the extent to which organisations are achieving their sustainable objectives (Hasan et al., 2024; Kumar et al., 2024; Shukla et al., 2024).

Table 1. Performance indicators.

No.	Performance indicator for Green Intelligence	Description	STEM contribution	Role of AI in implementation and monitoring
1	**Carbon emission reduction**	Measurement and reduction of CO_2 emissions from operational activities	STEM contributes by developing carbon capture technologies, renewable energy solutions, and energy efficiency methods	AI enables real-time emission monitoring, forecasting, and process optimization for reduction
2	**Digital resource efficiency**	Measures the efficiency of digital resource usage, including servers, storage, and computational processes	STEM supports the development of energy-efficient digital infrastructures and optimization technologies	AI optimises resource usage through data analysis and reduces consumption based on operational requirements
3	**Environmental impact**	Evaluate the effects of organisational activities on the environment, including pollution and waste	STEM supports recycling methods, pollution reduction, and efficient water and energy consumption	AI analyses environmental impact and provides forecasts to reduce waste and pollution through simulations and predictive models
4	**Adoption of green technologies**	Measures the implementation of eco-friendly technologies such as renewable energy and low-emission solutions	STEM contributes by developing green energy solutions and emission-reduction technologies	AI aids in selecting suitable technologies through cost-benefit analysis and long-term performance forecasting
5	**Energy efficiency**	Monitors energy consumption and reduces energy losses in organizational processes	STEM provides methods to improve energy efficiency, such as upgrading heating, ventilation and air conditioning (HVAC) systems and insulation	AI manages and monitors energy consumption in real time, minimising losses and optimizing energy distribution

AI and STEM skills play an important role in these assessments by automating complex processes and improving the accuracy of environmental impact measurements (Gueye and Exposito, 2020; Kumar et al., 2024; Rigotti and Fosch-Villaronga, 2024; Rodgers et al., 2023; UNESCO, 2021; World Economic Forum, 2023b). For example, AI can analyse resource consumption and emissions in a way that was not previously possible, while STEM technologies facilitate resource management and waste reduction innovations. Using advanced algorithms and data analysis methods, organisations can identify trends and optimise sustainability initiatives (Franki et al., 2023; Su et al., 2023).

To implement sustainable initiatives, organisations need special STEM skills to support this process. Technology and science competencies are essential for understanding and applying specific performance indicators such as emission reduction and resource management. Consequently, STEM-qualified employees can play a role in achieving sustainable transformation by improving organisational performance and environmental outcomes (Ahmed, 2016; Akgündüz et al., 2022; Australian Government, 2020; Kiazad et al., 2024).

4.3 The concept of Green Intelligence performance indicators

Performance indicators (KPIs) for Green Intelligence are measurable criteria that help organisations monitor and evaluate the impact of their green actions (Alzoubi and Mishra, 2024). As can be seen in Table 1, these indicators cover a wide range of aspects, including carbon reduction, digital resource efficiency, and environmental impact. KPIs provide a clear and concise methodology to measure the effectiveness of Green Intelligence initiatives and identify areas where there are opportunities for improvement. By tracking and evaluating these KPIs, companies can adjust their strategies and develop training programmes to reflect observed performance, ensuring continuity in achieving sustainability goals. International frameworks, such as the Global Reporting Initiative (GRI) and the Carbon Disclosure Project (CDP) (de Souza et al., 2019), play a key role in assessing and standardising sustainability reporting, providing organisations with a structured guide for measuring and reporting environmental impacts (Alzoubi and Mishra, 2024; Dieterle et al., 2022). These frameworks, which are internationally recognised, allow companies to align with international standards and implement verified sustainability practices. For example, the GRI includes specific guidelines for reporting emissions and energy consumption, providing a solid basis for the development of emission reduction initiatives.

At the same time, frameworks such as ISO 14001 and the Sustainability Accounting Standards Board (SASB) add a dimension of industrial assessment and specificity, bringing more rigour to the implementation and reporting of Green Intelligence (Parfitt, 2024).

However, using these frameworks is not without its challenges. They can be relatively complex and require adaptation to the specifics of each organisation. Here, AI and STEM skills offer a solution: By implementing AI-based analytics systems, organisations can optimise and customise these frameworks according to their needs.

These systems facilitate collection and analysis, thus transforming the reporting process into a mechanism for continuous improvement of green strategies.

4.4 The Role of STEM and AI in the evaluation and implementation of Green Intelligence initiatives

In recent years, emerging technologies based on AI and STEM skills have transformed the way organisations approach sustainability. STEM and AI provide essential support for collecting, interpreting, and using data on Green Intelligence performance indicators. Specifically, AI systems can process large volumes of data related to energy consumption and carbon emissions, providing organisations with a clear and up-to-date picture of their environmental performance. AI also enables the creation of predictive algorithms that help organisations anticipate and respond promptly to changes in the business environment, thus reducing environmental risks (Jeevaraj et al., 2023; Yu et al., 2023). In the evaluation of Green Intelligence initiatives, STEM is important to ensure that measurements are accurate and results are replicable and based on validated methodologies (Alzoubi and Mishra, 2024). Therefore, AI and STEM support and accelerate the transition to sustainable practices in organisations.

In the context of Green Intelligence initiatives, STEM education and AI are the fundamental pillars for the development of sustainable and innovative solutions (Dzhunushalieva and Teuber, 2024; Kirimtat et al., 2020). Their role transcends the simple integration of technology in the educational process, orienting themselves toward the evaluation and implementation of ecological strategies (Bibri, 2020; Lv et al., 2022). AI, applied within STEM, facilitates the creation of advanced models for optimising resources, monitoring the environment, and reducing climate impact, thus strengthening the foundation of a green economy. Therefore, the synergy between STEM and AI (Ng, 2019; Rashid and Kausik, 2024) becomes essential to anticipate and manage environmental challenges, redefining educational paradigms, providing students and specialists with the necessary skills to support the transition to sustainability (Fakayode et al., 2019; Kousloglou et al., 2023; Mirea et al., 2023; OECD, 2023).

4.5 Case Study: Stanford University's net zero campuses

A relevant example of the synergy between STEM education, AI, and Green Intelligence initiatives is the use of machine learning algorithms to optimise the energy consumption of smart university campuses. Stanford University (Chou and Nguyen, 2024; Marcone et al., 2023) has implemented an AI-based system to effectively monitor and manage the campus's energy consumption, with the goal of reducing its carbon footprint (Colella et al., 2005; Manzano-Agugliaro et al., 2023). This system integrates data from various sources, such as historical energy consumption, weather conditions, and building usage, to automatically adjust the operation of HVAC systems.

The role of STEM fields in the promotion of sustainability initiatives is demonstrated by their integration into Green Intelligence projects. ***Science*** plays an important role in the analysis of the effects of carbon emissions on the environment and in providing theoretical models to optimise resource use. ***Technology*** is contributing to the development of IoT sensors and digital infrastructure to collect and transmit real-time data to enable accurate monitoring and control. ***Engineering*** supports these efforts by creating AI algorithms to ensure sustainable construction operation, and ***Mathematics*** provides a basis for predictive modelling and energy resource optimisation. The results of these initiatives are remarkable. For example, in the first three years of implementation, energy consumption has been reduced by more than 30%, resulting in significant financial savings and improved operational sustainability.[2] Moreover, these efforts have produced a global replicable model that allows other institutions to move toward zero-emission buildings and contribute to broader sustainability goals (Bibri, 2019; Wright et al., 2022).

This approach is closely related to education, as it serves as a dynamic platform for students engaged in STEM research. It provides them with the opportunity to practice emerging technologies and understand how AI is applied to green initiatives. In addition, it promotes the development of future-critical thinking and the ability to deal with complex environmental challenges. Through this lens, Stanford's project highlights the transformational potential of STEM and AI in the promotion of Green Intelligence initiatives. Through the integration of cutting-edge technologies and sustainability-focused education, learning becomes a powerful catalyst for environmental management and innovation.

5. Future trends: stem-enabled workforce as a catalyst for Green Intelligence and sustainable innovation

In a world driven by rapid digital transformation and the aggressive climate crisis, the convergence of science, technology, and ecological sustainability has become a topic of academic and practical importance. Recent research (Li et al., 2024; Wang et al., 2023) has highlighted the essential role of AI in monitoring carbon dioxide emissions, resource optimisation, and green technology development. By 2023, more than 50% of the articles published in *Nature Sustainability*[3] include one or more of the 17 Sustainable Development Goals, reflecting the university's commitment to global issues (Dzhunushalieva and Teuber, 2024; *Journal Information | Nature Sustainability*, n.d.). This trend highlights the growing importance of sustainability in contemporary research, including areas such as green technologies, environmental policies, and ecological innovations (Du et al., 2024). Research ranging from the optimisation of natural resources to predictive models driven by AI shows that scientific research is strongly linked to the global sustainability agenda and supports the transition to a more sustainable economy (IEA, 2021; Köhler et al., 2019).

At the same time, STEM education plays an essential role in preparing workers capable of implementing sustainable solutions (Su and Guo, 2023). Emerging concepts such as *"green AI"* (Bolón-Canedo et al., 2024; Yigitcanlar et al., 2021), advocate the use of technology to advance ecological progress (Alzoubi and Mishra,

[2] https://sesi.stanford.edu/.

[3] https://www.nature.com/natsustain/journal-information.

2024). This synergy solves global challenges, drives innovation, and opens up new opportunities for industries and policies.

This convergence can also be described as an integrated platform that combines AI analysis and automation capabilities, multidisciplinary skills promoted through STEM education, and sustainable principles promoted by Green Intelligence to create technological solutions to global challenges.

5.1 The foundations of a new green concept

Theoretical and applied conceptual frameworks integrated with AI, STEM education, and Green Intelligence have been explored in the literature but are still lacking comprehensive development (Table 2). Recent studies have highlighted the tendency to use AI to manage green innovation processes, such as reducing carbon emissions by optimising resources and creating sustainable ecosystems within organisations (Belkhir and Elmeligi, 2018; IEA, 2021; Verdecchia et al., 2023). Furthermore, STEM education environments are highlighted as a training ground for sustainable workers. These results suggest that AI and an innovative organisational culture can strengthen the impact of STEM in the development of green solutions. Recent studies (Strielkowski et al., 2024) have confirmed, for example, the importance of models and knowledge sharing through AI to promote green innovations (Carayannis and Morawska-Jancelewicz, 2022; Yu and Zhang, 2024).

Table 2. New green-concept foundations.

No.	Theoretical concepts	Concepts foundations
1.	AI	Provides advanced prediction and analysis capabilities used to optimise industrial processes and develop more efficient energy systems. AI models the effects of climate change and proposes solutions to reduce carbon dioxide emissions through machine learning and evolutionary algorithms.
2.	STEM education	STEM equips the workforce with advanced technology management and innovation skills in emerging areas. Integrating STEM principles into education builds a strong foundation for understanding the relationship between technology and sustainability.
3.	STEM-enable workforce	STEM-enabled workers are groups of professionals whose knowledge and skills come from STEM education. These individuals are able to use their interdisciplinary expertise to solve complex problems, adapt to new technologies, and promote innovation and are essential to drive the development of green technologies, AI, sustainability, and economic growth, as well as to address global challenges such as climate change, resource management, and digital transformation.
4.	Green Intelligence	The focus is on responsible use of natural resources, reduction of pollution, and development of the circular economy. The fusion of technology and sustainability facilitates the transition to a green economy.

5.2 Synergies trends between AI, STEM and Green Intelligence

The intersection of AI, STEM, and Green Intelligence fosters a collaboration between three main components:

- *Technological innovation*: Science and technology provide core competencies; AI optimises processes, while Green Intelligence prioritises environmental sustainability and energy efficiency (Linkedin Economic Graph, 2022).
- *Sustainable design*: AI can analyse product life cycles, and STEM provides engineering methods to implement data-driven changes (Mirea et al., 2024).
- *Economic development*: Green technologies supported by STEM education create jobs in innovative areas such as renewable energy and precision agriculture (Ahmed, 2016; Let's Talk Science, 2019; Pompigna and Mauro, 2022; Tavares et al., 2023).

Our research identified three key applications:

- *Precision agriculture*: By using AI for soil analysis and STEM to develop autonomous drones, agricultural resources can be reduced and sustainable production can be supported (Bhat and Huang, 2021; Su et al., 2023; Zhang et al., 2021);
- *Renewable energy*: AI algorithms optimise energy grids and STEM develops more efficient solar panels (Ali and Choi, 2020; Bouquet et al., 2024; SaberiKamarposhti et al., 2024);
- *SMART cities*: AI and STEM contribute to the construction of low-emission cities, promoting green mobility and recycling waste (Courage, 2017; He et al., 2023; Lv et al., 2022; Peng et al., 2023; Statistical Office of European Communities, 2023; UrbanizeHub, 2024; World Economic Forum, 2023c, 2023a).

A conceptual theoretical framework that integrates AI, STEM, and Green Intelligence would be essential for trends in scientific research. We believe that an appropriate term to define the intersection point between AI, STEM education, and Green Intelligence could be "***Eco-Techno-Synergy***". It captures the idea of a synergy between advanced technology and sustainability, based on STEM education as the foundation for training a workforce capable of designing green solutions through the use of AI. This name emphasises the collaboration between human intelligence, technological innovation, and environmental responsibility, which is a central point of conclusions related to the global transition to sustainability. A review of the major academic databases, including Google Scholar and Web of Science, revealed that the term "***Eco-Techno-Synergy***" has not yet been recognized as an established concept in academic literature. It is an original formulation that combines two major concepts, technology and ecology, through the interdisciplinary collaboration facilitated by STEM education.

This theoretical conceptual framework is an innovative concept with academic and practical potential, which can develop the following implications (Table 3) and contribute to transforming the theory into solutions applicable in everyday life, from education to industry and public policy.

Table 3. Conceptual integration of AI, STEM, and Green Intelligence: research needs, practical applications, and real-world examples.

Category	**Key Points**	**Examples**
1. Necessity in research		
Interdisciplinarity	Facilitates connections across traditional fields, enabling AI and STEM to support the transition to sustainable practices.	Studies on the role in renewable energy forecasting and biodiversity conservation (Alzoubi and Mishra, 2024; Bolón-Canedo et al., 2024; Melnyk et al., 2019).
Hypothesis generation	Provides a foundation for empirical studies on the impact on sustainability and education.	Research exploring AI predictive capabilities in managing carbon emissions (Bibri, 2020; Bibri et al., 2023; Mora et al., 2023).
Green innovation validation	Contributes to the creation of standardised methods for assessing the environmental impact of technological solutions.	Development of frameworks to measure the carbon footprint of digital technologies (Jeevaraj et al., 2023; Maitra, 2024; Škare et al., 2024).
2. Practical utility		
Education and training	Applicable in STEM education programmes to prepare a workforce capable of developing green solutions.	The EU GreenComp framework integrates sustainability into STEM education curricula (Ilardo and Salinaro, 2024; Laherto et al., 2023; Sourgiadaki and Karkalakos, 2023; Valérie, 2024).
Organizational streamlining	Guides companies to integrate AI to reduce resource consumption and manage energy efficiently.	Amazon AI-powered energy management systems in warehouses (Gunasekar et al., 2023; Kumar et al., 2024).
Urban and ecological planning	Useful for designing smart cities that minimise ecological footprints.	Singapore's smart city initiatives improve energy use and waste management (Demirel, 2023).
3. Applications in sectors		
Urban sustainability	Cities like Singapore and Amsterdam use AI to improve energy use and manage resources in smart city projects.	Amsterdam's circular economy programme uses AI for waste management and recycling (Calisto Friant et al., 2023).
Industry	Tesla uses AI to improve production processes and reduce carbon footprint in the manufacture of electric vehicles.	Tesla's Gigafactory employs AI (AI) for energy-efficient operations (Henrik Simon Schöttle, 2024; Mercante Thierauf, 2024).
Education	STEM initiatives in the US and the EU incorporate sustainability and AI modules in school programmes.	STEM curricula in Finland integrate coding, sustainability and environmental awareness (Mäkelä et al., 2023; Marzuki et al., 2024).
Precision agriculture	Platforms such as Climate Corporation use AI to improve crop yields with minimal environmental impact.	Climate Corporation AI predicts weather patterns to improve farming efficiency (Pachiappan et al., 2023; Rashid and Kausik, 2024).

All these initiatives are practical steps towards a more complex model that completely unifies the three areas. Trends lead to the intersection, which can be defined as "***Eco-Driven Techno-Workforce***", a new concept that reflects a STEM-educated workforce that uses AI and technological skills to develop sustainable solutions, reducing the environmental impact of industry, and promoting green innovation.

This intersection combines the following.

- **STEM:** Training analytical, technical, and innovation skills.
- **AI:** Using algorithms to streamline processes.
- **GREEN INTELLIGENCE:** Directing these resources towards ecological and sustainable objectives.

It is a bridge between education, technology, and environmental responsibility.

Both concepts "***Eco-Techno-Synergy***" and "***Eco-Driven Techno-Workforce***," although connected, are not identical and can be approached complementary, leading to the enrichment of the perspectives offered (Table 4).

This comparison highlights the complexity of the relationship between education, technology, and ecology and provides a progressive perspective, from theory to application, attracting readers interested in both concepts and practical examples. It also allows for discussion of the challenges and gaps in the implementation of the two concepts. Global challenges require innovative and sustainable solutions, and the integration of STEM education, AI, and Green Intelligence is an innovative step. The workforce equipped with STEM is unique in its role as a catalyst, enabling advanced technologies to bring ecological responsibility together. Future trends show that investing in these workers not only promotes green innovation, but also ensures a sustainable balance between technological progress and environmental management and shapes industries, policies, and societies.

Table 4. Comparing ***Eco-Techno-Synergy*** and ***Eco-Driven Techno-Workforce***: conceptual versus applied frameworks.

Aspect	**Eco-Techno-Synergy**	**Eco-Driven Techno-Workforce**
1. Conceptual focus	Broadly conceptual, focussing on the interplay between AI, STEM, and Green Intelligence in systems and processes.	Applied focus on the role of the STEM-trained workforce in implementing green solutions.
2. Primary goal	Harmonising technology and ecology at the systemic and process levels.	Preparing the workforce to translate conceptual frameworks into actionable results.
3. Scope	Relevant for global policies, infrastructures, and interdisciplinary collaborations.	Focused on education, professional innovation and localised industrial impact.
4. Integration approach	Provides the theoretical context and the interdisciplinary framework.	Details practical implementation, showcasing how human factors drive sustainability.

6. Conclusions

The integration of Green Intelligence, STEM skills, and AI offers transformative potential for the development, sustainability, and innovation of the workforce. This study has demonstrated how these elements, when synergised, create a powerful framework to address the global challenges of climate change, resource optimisation, and ecological resilience. By equipping the workforce with the knowledge, skills, and competencies needed to navigate this complex landscape, organisations can not only meet their environmental objectives but also drive long-term economic growth and competitiveness.

Green Intelligence, as a concept, emphasizes the collective capacity of a workforce to understand and implement sustainable practices while adapting to evolving environmental demands. This extends beyond specialised green roles to encompass general workforce capabilities, embedding sustainability into everyday business operations. STEM skills, on the other hand, serve as the foundation for innovation, enabling individuals to design, implement, and scale technologies that address environmental, social, and economic imperatives. The interdisciplinary approach of STEM fosters critical thinking and problem solving, essential for advancing sustainability initiatives. (Glick et al., 2024; Odugbesan et al., 2022; Olatunbosun Bartholomew Joseph and Nwankwo Charles Uzondu, 2024; WorkDove, 2024)

The role of AI in improving workforce capabilities is equally significant. AI-powered tools and methodologies enable personalised learning experiences, adaptive assessments, and real-time feedback mechanisms, accelerating the acquisition of critical skills. Additionally, AI improves decision making by analyzing complex datasets to inform sustainable strategies, optimise processes, and minimizing resource use.

However, the practical implementation of this framework is not without challenges. The persistent gap between theoretical models and real-world applications, particularly in aligning educational outcomes with industry needs, requires targeted strategies. (Fatol and Draghici, 2023; Fatol and Robescu, 2023) Collaborative partnerships between academia and business are essential to bridge this divide, ensuring that educational curricula reflect the evolving demands of a green economy.

This study highlights the need for continued investment in workforce sustainability, prioritising the integration of Green Intelligence, STEM education, and AI technologies. Future research should focus on refining implementation frameworks, assessing long-term impacts, and addressing challenges such as skills gaps and ethical considerations in AI-driven solutions. By adopting these approaches, organisations can build a resilient workforce equipped to lead the transition toward a sustainable future.

References

Ahmed, H. O. K. (2016). Strategic future directions for developing STEM education in higher education in Egypt as a driver of the innovation economy. Journal of Education and Practice 7(8): 127–145.

Akgündüz, D., Akgündüz, D., Topalsan, A. K. and Turk, Z. (2022). The views of academicians and specialists on STEM and related concepts: Research Article. Journal of Turkish Science Education 19(2): 438–464. https://doi.org/10.36681/.

Alamäki, A., Nyberg, C., Kimberley, A. and Salonen, A. O. (2024). Artificial Intelligence literacy in sustainable development: A learning experiment in higher education. Frontiers in Education 9: 1343406. https://doi.org/10.3389/FEDUC.2024.1343406/BIBTEX.

Ali, S. S. and Choi, B. J. (2020). State-of-the-art Artificial Intelligence techniques for distributed smart grids: A review. Electronics (Switzerland) 9(6): 1–28. https://doi.org/10.3390/electronics9061030.

Alzoubi, Y. I. and Mishra, A. (2024). Green Artificial Intelligence initiatives: Potentials and challenges. Journal of Cleaner Production 468: 143090. https://doi.org/10.1016/J.JCLEPRO.2024.143090.

Australian Government. (2020). STEM Equity Monitor - Summary Report 2020. 12.

Beamery. (2024a). Bridging The Green Skills Gap: A Strategy For Building A Sustainable Workforce. https://beamery.com/resources/blogs/bridging-the-green-skills-gap-a-strategy-for-building-a-sustainable-workforce.

Beamery. (2024b). Talent Intelligence Explained: Unlocking The Power of Data In HR. https://beamery.com/resources/blogs/what-is-talent-intelligence.

Becker, Dr. N. (2024). STEM Skills and Their Role in Solving Global Sustainability Challenges [2024] - ClimateSort. https://climatesort.com/stem-skills/.

Belkhir, L. and Elmeligi, A. (2018). Assessing ICT global emissions footprint: Trends to 2040 & recommendations. Journal of Cleaner Production 177: 448–463. https://doi.org/10.1016/j.jclepro.2017.12.239.

Bhat, S. A. and Huang, N. F. (2021). Big Data and AI revolution in precision agriculture: survey and challenges. IEEE Access 9: 110209–110222. https://doi.org/10.1109/ACCESS.2021.3102227.

Bibri, S. E. (2019). The anatomy of the data-driven smart sustainable city: instrumentation, datafication, computerisation and related applications. Journal of Big Data 6(1): 59. https://doi.org/10.1186/s40537-019-0221-4.

Bibri, S. E. (2020). The eco-city and its core environmental dimension of sustainability: green energy technologies and their integration with data-driven smart solutions. Energy Informatics 3(1): 4. https://doi.org/10.1186/s42162-020-00107-7.

Bibri, S. E., Alexandre, A., Sharifi, A. and Krogstie, J. (2023). Environmentally sustainable smart cities and their converging AI, IoT, and big data technologies and solutions: an integrated approach to an extensive literature review. In Energy Informatics (Vol. 6, Issue 1, p. 9). Springer Nature. https://doi.org/10.1186/s42162-023-00259-2.

Bolón-Canedo, V., Morán-Fernández, L., Cancela, B. and Alonso-Betanzos, A. (2024). A review of green Artificial Intelligence: Towards a more sustainable future. Neurocomputing 599: 128096. https://doi.org/10.1016/j.neucom.2024.128096.

Borgonovi, F., Calvino, F., Criscuolo, C., Nania, J., Nitschke, J., O'Kane, L. et al. (2023). Emerging trends in AI skill demand across 14 OECD countries. In OECD Artificial Intelligence Papers (Vol. 2, Issue 2). https://doi.org/10.1787/7C691B9A-EN.

Bouquet, P., Jackson, I., Nick, M. and Kaboli, A. (2024). AI-based forecasting for optimised solar energy management and smart grid efficiency. International Journal of Production Research 62(13): 4623–4644. https://doi.org/10.1080/00207543.2023.2269565.

Calisto Friant, M., Reid, K., Boesler, P., Vermeulen, W. J. V. and Salomone, R. (2023). Sustainable circular cities? Analysing urban circular economy policies in Amsterdam, Glasgow, and Copenhagen. Local Environment 28(10): 1331–1369. https://doi.org/10.1080/13549839.2023.2206643.

Carayannis, E. G. and Morawska-Jancelewicz, J. (2022). The futures of Europe: Society 5.0 and Industry 5.0 as driving forces of future universities. Journal of the Knowledge Economy 13(4): 3445–3471. https://doi.org/10.1007/s13132-021-00854-2.

Chen, L., Chen, Z., Zhang, Y., Liu, Y., Osman, A. I., Farghali, M. et al. (2023). Artificial Intelligence-based solutions for climate change: a review. pp. 2525–2557. *In*: Environmental Chemistry Letters (Vol. 21, Issue 5). Springer. https://doi.org/10.1007/s10311-023-01617-y.

Cheng, C., Ahmad, S. F., Irshad, M., Alsanie, G., Khan, Y., Ahmad, A. Y. A. B. et al. (2023). Impact of green process innovation and productivity on sustainability: the moderating role of environmental awareness. Sustainability 15(17): 12945. https://doi.org/10.3390/SU151712945.

Cheng, J., Mahinder Singh, H. S., Zhang, Y. C. and Wang, S. Y. (2023). The impact of business intelligence, big data analytics capability, and green knowledge management on sustainability performance. Journal of Cleaner Production 429: 139410. https://doi.org/10.1016/j.jclepro.2023.139410.

Chiangpradit, L. (2024). STEM & Sustainability: Preparing the Next Generation for Green Careers - STEM Sports. https://stemsports.com/stem-sustainability-preparing-the-next-generation-for-green-careers/.

Chou, J. S. and Nguyen, H. M. (2024). Simulating long-term energy consumption prediction in campus buildings through enhanced data augmentation and metaheuristic-optimized Artificial Intelligence. Energy and Buildings 312: 114191. https://doi.org/10.1016/J.ENBUILD.2024.114191.

Colella, W. G., Niemoth, C. R., Lim, C. Y. and Hein, A. P. (2005). Evaluation of the financial and environmental feasibility of a network of distributed 200 kWe combined heat and power fuel cell systems on the stanford university campus. Fuel Cells 5(1): 148–166. https://doi.org/10.1002/fuce.200400074.

Courage, C. (2017). Arts in place: The arts, the urban and social practice. Arts in Place: The Arts, the Urban and Social Practice 1–228. https://doi.org/10.4324/9781315659299.

de Souza, J. T., de Francisco, A. C., Piekarski, C. M. and do Prado, G. F. (2019). Data mining and machine learning to promote smart cities: A systematic review from 2000 to 2018. In Sustainability (Switzerland) (Vol. 11, Issue 4). MDPI. https://doi.org/10.3390/su11041077.

Deloitte, UK. (2024). Green skills for the green economy Report. https://www.deloitte.com/uk/en/services/consulting/research/green-skills-for-green-economy.html.

Demirel, D. (2023). The impact of managing diversity on building the smart city a comparison of smart city strategies: cases from Europe, America, and Asia. SAGE Open 13(3): 1–18. https://doi.org/10.1177/21582440231184971.

Dieterle, E., Dede, C. and Walker, M. (2022). The cyclical ethical effects of using Artificial Intelligence in education. AI and Society. https://doi.org/10.1007/s00146-022-01497-w.

Du, J., Cai, H. and Jin, X. (2024). Exploring the association between artificial intelligence management and green innovation: expanding the research field for sustainable outcomes. Sustainability (Switzerland) 16(21): 9315. https://doi.org/10.3390/su16219315.

Duterte, J. P. (2024). Integrating AI-powered tools in stem education: assessing the impact of chatgpt on pedagogical practices in higher education institutions across ASEAN. EPRA International Journal of Research and Development (IJRD). https://doi.org/10.36713/epra2016.

Dzhunushalieva, G. and Teuber, R. (2024). Roles of innovation in achieving the Sustainable Development Goals: A bibliometric analysis. Journal of Innovation and Knowledge 9(2). https://doi.org/10.1016/j.jik.2024.100472.

Fakayode, S. O., Davis, J. J., Yu, L., Meikle, P. A., Darbeau, R., & Hale, G. (2019). Transforming stem departments for inclusion: creative innovation, challenges, adaptation, and sustainability at the university of arkansas-fort smith. https://doi.org/10.1108/S1479-364420190000022004.

Falebita, O. S. and Kok, P. J. (2024). Strategic goals for Artificial Intelligence integration among STEM academics and undergraduates in African higher education: a systematic review. Discover Education 3(1): 1–22. https://doi.org/10.1007/S44217-024-00252-1.

Fatol, D. and Draghici, A. (2023). Talent Management in the Context of Romanian Higher Education.

Fatol, D. and Robescu, D. (2023). Two worlds apart? overcoming the discord in talent management between academia and business. Acta Technica Napocensis 66.

Fatol, D., Robescu, D., Farkas, Z. and Draghici, A. (2020). Talent Management - Present and Future. Expanding Horizons Business, Management and Technology for Better Society.

Franki, V., Majnarić, D. and Višković, A. (2023). A comprehensive review of artificial intelligence (AI) companies in the power sector. In Energies (Vol. 16, Issue 3, p. 1077). Multidisciplinary Digital Publishing Institute. https://doi.org/10.3390/en16031077.

Glick, D., Miedijensky, S. and Zhang, H. (2024). Examining the effect of AI-powered virtual-human training on STEM majors' self-regulated learning behavior. Frontiers in Education 9: 1465207. https://doi.org/10.3389/FEDUC.2024.1465207.

Gueye, M. and Exposito, E. (2020). University 4.0 : The Industry 4.0 paradigm applied to Education. IX Congreso Nacional de Tecnologías En La Educación, January, 15.

Gunasekar, S., Ray, S., Dixit, S. K. and PA, M. R. (2023). AI-enables product purchase on Amazon: what are the consumers saying? Foresight 25(2): 185–193. https://doi.org/10.1108/FS-10-2021-0212.

Hasan, M. R., Khan, M. A. and Wuest, T. (2024). Towards Industry 5.0: A systematic literature review on sustainable and green composite materials supply chains.

He, X., Zhang, R., Yuan, X., Cao, Y. and Zhou, C. (2023). The role of planning policy in the evolution of the spatial structure of the Guangzhou metropolitan area in China. Cities 137: 104284. https://doi.org/10.1016/j.cities.2023.104284.

Henrik Simon Schöttle. (2024). Are Tesla's Gigafactory and autonomous driving technology, automotive industry disruptors? Universidade Catolica Portuguesa.

Iberdrola. (2024). Training and talent development - Iberdrola. https://www.iberdrola.com/careers/life-at-iberdrola/formation-training.

IEA. (2021). Net Zero by 2050: A Roadmap for the Global Energy Sector - International Energy Agency. International Energy Agency, 224.

Ilardo, M. and Salinaro, M. (2024). The GreenComp (the European Sustainability Competence Framework): a tool to reflect on social and environmental issues from an interdisciplinary perspective. 11–12. https://doi.org/10.6092/UNIBO/AMSACTA/7724.

Jatana, N., Wadhwa, D., Singh, N. K., Hassen, O. A., Gupta, C., Darwish, S. M. et al. (2024). Future frame prediction using generative adversarial networks. Karbala International Journal of Modern Science 10(1): 19–30. https://doi.org/10.33640/2405-609X.3338.

Jeevaraj, S., Gokasar, I., Deveci, M., Delen, D., Zaidan, B. B., Wen, X. et al. (2023). Adoption of energy consumption in urban mobility considering digital carbon footprint: A two-phase interval-valued Fermatean fuzzy dominance methodology. Engineering Applications of Artificial Intelligence 126: 106836. https://doi.org/10.1016/j.engappai.2023.106836.

Journal Information | Nature Sustainability. (n.d.). Retrieved November 23, 2024, from https://www.nature.com/natsustain/journal-information.

Khan, Y., Hassan, T., Tufail, M., Marie, M., Imran, M. and Xiuqin, Z. (2022). The nexus between CO_2 emissions, human capital, technology transfer, and renewable energy: evidence from Belt and Road countries. Environmental Science and Pollution Research. https://doi.org/10.1007/s11356-022-20020-8.

Khawaji, T. M. (2024). Best practices for implementing AI in STEM education: a systematic literature review. IJCSNS International Journal of Computer Science and Network Security 24(8). https://doi.org/10.22937/IJCSNS.2024.24.8.1.

Kiazad, K., Restubog, S. L. D., Hom, P. W., Capezio, A., Holtom, B. and Lee, T. (2024). STEMming the tide: New perspectives on careers and turnover. Journal of Organizational Behavior 45(3): 335–343. https://doi.org/10.1002/job.2780.

Kirimtat, A., Krejcar, O., Kertesz, A. and Tasgetiren, M. F. (2020). Future trends and current state of smart city concepts: a survey. pp. 86448–86467. In IEEE Access (Vol. 8). Institute of Electrical and Electronics Engineers Inc. https://doi.org/10.1109/ACCESS.2020.2992441.

Köhler, J., Geels, F. W., Kern, F., Markard, J., Onsongo, E., Wieczorek, A. et al. (2019). An agenda for sustainability transitions research: State of the art and future directions. Environmental Innovation and Societal Transitions 31: 1–32. https://doi.org/10.1016/j.eist.2019.01.004.

Kousloglou, M., Petridou, E., Molohidis, A. and Hatzikraniotis, E. (2023). Assessing students' awareness of 4Cs skills after mobile-technology-supported inquiry-based learning. Sustainability (Switzerland) 15(8): 6725. https://doi.org/10.3390/su15086725.

Kumar, V., Ashraf, A. R. and Nadeem, W. (2024). AI-powered marketing: What, where, and how? International Journal of Information Management 77: 102783. https://doi.org/10.1016/j.ijinfomgt.2024.102783.

Laherto, A., Rasa, T., Miani, L., Levrini, O. and Erduran, S. (2023). Future-oriented science education building sustainability competences: an approach to the European GreenComp framework. pp. 83–105. In Science Curriculum for the Anthropocene, Volume 2. Palgrave Macmillan, Cham. https://doi.org/10.1007/978-3-031-37391-6_5.

Let's Talk Science. (2019). Spotlight on science learning: A review of recent international and Canadian policy recommendations extended version.

Li, G., Luo, T., Liu, R., Song, C., Zhao, C., Wu, S. et al. (2024). Integration of carbon dioxide removal (CDR) technology and artificial intelligence (AI) in energy system optimization. Processes 12(2): 402. https://doi.org/10.3390/PR12020402.

Linkedin Economic Graph. (2022). Global Green Skills Report 2023. 9–9.

LinkedIn Talent Solutions. (2024). Green reports and resources. https://economicgraph.linkedin.com/research/green-skills-resources.

Lv, L., Zhang, S., Zhu, J., Wang, Z., Wang, Z., Li, G. et al. (2022). Ecological restoration strategies for mountainous cities based on ecological security patterns and circuit theory: a case of central Urban Areas in Chongqing, China. International Journal of Environmental Research and Public Health 19(24): 16505. https://doi.org/10.3390/ijerph192416505.

Maitra, R. (2024). Reducing carbon footprints through green technology in the tourism industry: a review of selected hotels in Delhi. Sustainable Tourism, Part B 105–121. https://doi.org/10.1108/978-1-83608-326-920241006.

Mäkelä, T., Tuhkala, A., Mäki-Kuutti, M. and Rautopuro, J. (2023). Enablers and constraints of STEM programme implementation: an external change agent perspective from a national STEM Programme in Finland. International Journal of Science and Mathematics Education 21(3): 969–991. https://doi.org/10.1007/s10763-022-10271-9.

Manzano-Agugliaro, F., Chihib, M., Chourak, M., Martínez, J. A., Zapata-Sierra, A. J. and Alcayde, A. (2023). Monitoring energy consumption of vending machines in university buildings. Energy Reports 10: 3252–3262. https://doi.org/10.1016/J.EGYR.2023.09.177.

Marcone, A. L., Darmstadt, G. L., Challamel, G. A., Mathur, M. B. and Gardner, C. D. (2023). Effects of an educational planetary plate graphic on meat consumption in a Stanford University dining hall: a randomized controlled trial. BMC Nutrition 9(1): 1–8. https://doi.org/10.1186/S40795-023-00764-3/TABLES/1.

Marzuki, O. F., Lih, E. T. Y., Zainol @ Abdullah, W. N. Z., Khairuddin, N., Inai, N. H., Md Saad, J. B. et al. (2024). Innovating education: a comprehensive review of STEM education approaches. International Journal of Academic Research in Progressive Education and Development 13(1). https://doi.org/10.6007/ijarped/v13-i1/20490.

Mazhar, T., Irfan, H. M., Haq, I., Ullah, I., Ashraf, M., Shloul, T. Al et al. (2023). Analysis of challenges and solutions of IoT in smart grids using AI and machine learning techniques: A review. Electronics (Switzerland): 12(1): 242. https://doi.org/10.3390/electronics12010242.

Melnyk, L., Kubatko, O., Dehtyarova, I., Matsenko, O. and Rozhko, O. (2019). The effect of industrial revolutions on the transformation of social and economic systems. Problems and Perspectives in Management 17(4): 381–391. https://doi.org/10.21511/ppm.17(4).2019.31.

Mercante Thierauf, T. (2024). Fixing the Future? The Controversy Surrounding Tesla's 'Gigafactory Berlin-Brandenburg' as a site of contested future-making in times of climate change. Time and Society. https://doi.org/10.1177/0961463X241273786.

Mercer. (2024). 2024 Global Talent Trends. https://www.mercer.com/insights/people-strategy/future-of-work/global-talent-trends/.

Mirea, N., Draghici, A. and Baesu, V. (2023). Economic, social and environmental sustainability: The role of technology and political dialogue a cause-effect analysis of entrepreneurship and STEM education. Management, Knowledge and Learning 525–532.

Mirea, N., Palade, M. and Gaureanu, A. (2024). STEM education, artificial intelligence, and ethical challenges. MakeLearn 251–260.

Mora, L., Gerli, P., Ardito, L. and Messeni Petruzzelli, A. (2023). Smart city governance from an innovation management perspective: Theoretical framing, review of current practices, and future research agenda. Technovation 123: 102717. https://doi.org/10.1016/j.technovation.2023.102717.

Ng, S. B. (2019). Exploring STEM competences for the 21st century. In-Progress Reflection 30: 1–53.

Odugbesan, T., Aghazadeh, S., Al Qaralleh, R. E. and Sogeke, O. S. (2022). Green talent management and employees' innovative work behavior: the roles of Artificial Intelligence and transformational leadership | Request PDF. https://www.researchgate.net/publication/360263270_Green_talent_management_and_employees'_innovative_work_behavior_the_roles_of_artificial_intelligence_and_transformational_leadership.

OECD. (2023). Teaching for the Future: Global Engagement, Sustainability and Digital Skills.

OECD. (2024). Skills for the green transition | OECD Employment Outlook 2024: The Net-Zero Transition and the Labour Market. https://www.oecd-ilibrary.org/employment/oecd-employment-outlook-2024_869462a8-en.

Olatunbosun Bartholomew Joseph and Nwankwo Charles Uzondu. (2024). Integrating AI and machine learning in STEM education: Challenges and opportunities. Computer Science & IT Research Journal 5(8): 1732–1750. https://doi.org/10.51594/csitrj.v5i8.1379.

Pachiappan, K., Anitha, K., Pitchai, R., Sangeetha, S., Satyanarayana, T. V. V. and Boopathi, S. (2023). Intelligent machines, IoT, and AI in revolutionizing agriculture for water processing. pp. 374–399. In Handbook of Research on AI and ML for Intelligent Machines and Systems. IGI Global. https://doi.org/10.4018/978-1-6684-9999-3.ch015.

Parfitt, C. (2024). Accounting for ethics: SASB and Integrated Reporting. pp. 94–116. In False Profits of Ethical Capital. Manchester University Press. https://doi.org/10.7765/9781526174239.00012.

Peng, H., Li, X., Yang, T. and Tan, S. (2023). Research on the relationship between the environmental characteristics of pocket parks and young people's perception of the restorative effects—a case study based on Chongqing City, China. Sustainability (Switzerland) 15(5): 3943. https://doi.org/10.3390/su15053943.

Pompigna, A. and Mauro, R. (2022). Smart roads: A state of the art of highways innovations in the Smart Age. Engineering Science and Technology, an International Journal 25: 100986. https://doi.org/10.1016/J.JESTCH.2021.04.005.

Portillo-Blanco, A., Deprez, H., De Cock, M., Guisasola, J. and Zuza, K. (2024). A systematic literature review of integrated STEM education: uncovering consensus and diversity in principles and characteristics. In Education Sciences (Vol. 14, Issue 9). Multidisciplinary Digital Publishing Institute (MDPI). https://doi.org/10.3390/educsci14091028.

Raman, R., Pattnaik, D., Lathabai, H. H., Kumar, C., Govindan, K. and Nedungadi, P. (2024). Green and sustainable AI research: an integrated thematic and topic modeling analysis. Journal of Big Data 11(1). https://doi.org/10.1186/s40537-024-00920-x.

Rashid, A. Bin and Kausik, M. A. K. (2024). AI revolutionizing industries worldwide: A comprehensive overview of its diverse applications. Hybrid Advances 7: 100277. https://doi.org/10.1016/j.hybadv.2024.100277.

Rigotti, C. and Fosch-Villaronga, E. (2024). Fairness, AI & recruitment. Computer Law and Security Review 53. https://doi.org/10.1016/j.clsr.2024.105966.

Rocki, B. (2024). Your Handy Guide to Talent Development Benchmarking. https://www.reworked.co/talent-management/talent-development-benchmarking-a-step-by-step-guide/.

Rodgers, W., Murray, J. M., Stefanidis, A., Degbey, W. Y. and Tarba, S. Y. (2023). An Artificial Intelligence algorithmic approach to ethical decision-making in human resource management processes. Human Resource Management Review 33(1). https://doi.org/10.1016/J.HRMR.2022.100925.

SaberiKamarposhti, M., Kamyab, H., Krishnan, S., Yusuf, M., Rezania, S., Chelliapan, S. et al. (2024). A comprehensive review of AI-enhanced smart grid integration for hydrogen energy: Advances, challenges, and future prospects. International Journal of Hydrogen Energy 67: 1009–1025. https://doi.org/10.1016/j.ijhydene.2024.01.129.

Sarfraz, M., Han, H., Ivascu, L., Ozturk, I. and Raza, M. A. (2024). Service robot performance and hospitality employees' job loss insecurity and turnover*. Innovation: The European Journal of Social Science Research. https://doi.org/10.1080/13511610.2024.2417932.

Savon, I., Ukraintsev, V. and Kanaki, V. (2020). Technological audit in the system of tools for assessing the innovative potential of technology transfer. IOP Conference Series: Materials Science and Engineering 709(4). https://doi.org/10.1088/1757-899X/709/4/044060.

Scientix. (2024). What is the Importance of STEM Education in Climate Change and Sustainability Education for Children? - Scientix blog Sustainable Development Goals Scientix blog. https://blog.scientix.eu/2024/06/what-is-the-importance-of-stem-education-in-climate-change-and-sustainable-development-goals-for-children/.

Shukla, V., Sant, A., Sharma, P., Nayak, M. and Khatri, H. (2024). An explainable Artificial Intelligence based approach for the prediction of key performance indicators for 1 megawatt solar plant under local steppe climate conditions. Engineering Applications of Artificial Intelligence 131: 107809. https://doi.org/10.1016/j.engappai.2023.107809.

Simmonds, P. and Lally, C. (2024). Green skills in education and employment.

Škare, M., Gavurova, B. and Porada-Rochon, M. (2024). Digitalization and carbon footprint: Building a path to a sustainable economic growth. Technological Forecasting and Social Change 199: 123045. https://doi.org/10.1016/j.techfore.2023.123045.

Sourgiadaki, M. and Karkalakos, S. (2023). "GreenComp" as a tool for examining motivation of vocational teachers to create learning opportunities for the green transition. SN Social Sciences 3(7): 1–27. https://doi.org/10.1007/s43545-023-00699-3.

Statistical Office of the European Communities. (2023). Sustainable development in the European Union Monitoring report on progress towards the SDGs in an EU context 2023 edition. In Publications Office of the European Union.

Strielkowski, W., Kuzmin, E. and Vlasov, M. (2024). Integrative approaches across green energy technologies: the role of innovations in human capital. Academia Green Energy. https://doi.org/10.20935/acadenergy6244.

Su, J., Zhu, X., Li, S. and Chen, W. H. (2023). AI meets UAVs: A survey on AI empowered UAV perception systems for precision agriculture. pp. 242–270. In Neurocomputing (Vol. 518). Elsevier. https://doi.org/10.1016/j.neucom.2022.11.020.

Su, Q. and Guo, S. (2023). STEM-based principles and strategies to cultivate students' social and emotional learning. STEM Education Review 1. https://doi.org/10.54844/STEMER.2023.0374.

Talent Management Institute. (2024). Sustainable Workforce Development: Talent Strategies for the Future. https://www.tmi.org/blogs/sustainable-workforce-development-talent-strategies-for-the-future.

Tavares, M. C., Azevedo, G., Marques, R. P. and Bastos, M. A. (2023). Challenges of education in the accounting profession in the Era 5.0: A systematic review. In Cogent Business and Management (Vol. 10, Issue 2). Cogent. https://doi.org/10.1080/23311975.2023.2220198.

Telfort, R. (2024). Artificial Intelligence in STEM Higher Education: Opportunities, Challenges & Ethical Dilemmas. https://www.norc.org/research/library/artificial-intelligence-in-stem-higher-education.html.

Tornjanski, V., Snežana Knežević, Mirčetić, V., Drinkwater, K., Alzoubi, H. M., Juraev, D. A. et al. (2024). The role of green society in Society 5.0: Tango diamond in a collective intelligence (Hybrid) ecosystem founded on human-centricity and sustainability. Journal of Business and Social Sciences.

UNESCO. (2021). AI and education Guidance for policymakers.

UNESCO. (2024a). Ethics of Artificial Intelligence. https://www.unesco.org/en/artificial-intelligence/recommendation-ethics.

UNESCO. (2024b). What you need to know about UNESCO's new AI competency frameworks for students and teachers. https://www.unesco.org/en/articles/what-you-need-know-about-unescos-new-ai-competency-frameworks-students-and-teachers.

UrbanizeHub. (2024). Analiză: Topul celor mai inspiraționale orașe din 2024 - UrbanizeHub. https://urbanizehub.ro/analiza-topul-celor-mai-inspirationale-orase-din-2024/.

Valdes-Ramirez, D., de Armas Jacomino, L., Monroy, R. and Zavala, G. (2024). Assessing sustainability competencies in contemporary STEM higher education: a data-driven analysis at Tecnologico de Monterrey. Frontiers in Education 9: 1415755. https://doi.org/10.3389/FEDUC.2024.1415755/BIBTEX.

Valérie, A. (2024). GreenComp in practice: case studies on the use of the European. <bound method Organization.get_name_with_acronym of <Organization: Technopolis Group>>.

Verdecchia, R., Sallou, J. and Cruz, L. (2023). A systematic review of Green AI. In Wiley Interdisciplinary Reviews: Data Mining and Knowledge Discovery (Vol. 13, Issue 4, p. e1507). John Wiley & Sons, Ltd. https://doi.org/10.1002/widm.1507.

Wang, H., Fu, T., Du, Y., Gao, W., Huang, K., Liu, Z. et al. (2023). Scientific discovery in the age of Artificial Intelligence. Nature 620(7972): 47–60. https://doi.org/10.1038/S41586-023-06221-2.

WorkDove. (2024). Talent Development Basics: Definition, Examples, and Strategies. https://workdove.com/talent-development-definition/.

World Economic Forum. (2023a). Cybersecurity Futures 2030: New Foundations | World Economic Forum.

World Economic Forum. (2023b). M A Y 2 0 2 3/the-future-of-jobs-report-2023.

World Economic Forum. (2023c). World Economic Forum Global Gender Gap Index. https://www.weforum.org/search/?query=smart+city.

World Economic Forum. (2024). Green jobs grow twice as fast as workers with green skills | World Economic Forum. https://www.weforum.org/stories/2024/02/green-jobs-green-skills-growth/.
Wright, J., Ackley, R., Gopal, S. and Phillips, N. (2022). The BosWash Infrastructure Biome and Energy System Succession. Infrastructures 7(7): 95. https://doi.org/10.3390/infrastructures7070095.
Yigitcanlar, T., Mehmood, R. and Corchado, J. M. (2021). Green Artificial Intelligence: towards an efficient, sustainable and equitable technology for smart cities and futures. Sustainability (Switzerland) 13(16): 8952. https://doi.org/10.3390/su13168952.
Yin, H. T., Wen, J. and Chang, C. P. (2023). Going green with Artificial Intelligence: The path of technological change towards the renewable energy transition. Oeconomia Copernicana 14(4): 1059–1095. https://doi.org/10.24136/OC.2023.032.
Yu, H., Wang, J., Hou, J., Yu, B. and Pan, Y. (2023). The effect of economic growth pressure on green technology innovation: Do environmental regulation, government support, and financial development matter? Journal of Environmental Management 330: 117172. https://doi.org/10.1016/j.jenvman.2022.117172.
Yu, L. and Zhang, Q. (2024). Measurement of new qualitative productivity kinetic energy from the perspective of digital and green collaboration—comparative study based on European countries. Journal of Cleaner Production 476: 143787. https://doi.org/10.1016/J.JCLEPRO.2024.143787.
Yu, Z., Zong, Z., Zheng, W. and Ma, C. (2024). Greening the Workplace: Examining the Relationship Between Green Talent Management and Pro-Environmental Behaviors in Chinese Public Sectors. Https://Doi.Org/10.1177/00910260241253391. https://doi.org/10.1177/00910260241253391.
Zhang, P., Guo, Z., Ullah, S., Melagraki, G., Afantitis, A. and Lynch, I. (2021). Nanotechnology and Artificial Intelligence to enable sustainable and precision agriculture. pp. 864–876. In Nature Plants (Vol. 7, Issue 7). Nature Publishing Group. https://doi.org/10.1038/s41477-021-00946-6.

Chapter 7

Digital Transformation in Fashion

Advancing to a Sustainable ZW Based Industry

Emil-Constantin Loghin, Irina-Elena Marin,* Manuela-Lăcrămioara Avădanei, Ion Verzea and Ionuț Dulgheriu*

1. Introduction

The main goal of the manufacturing companies in the fashion industry is to shift from mass production to mass customization production (Newman, 2017). The trend towards mass customization has offered manufacturers the advantage of responding more effectively to the customers' demand (Newman, 2017). Thus, the societal benefits of digital transformation aim to improve people's living standards by providing highly customized products, quality customized products and a better work environment (Wang et al., 2016). In addition, some of the benefits for the employees are improved human learning through the use of intelligent assistance, thus increasing motivation and job satisfaction (Müller et al., 2018). Changes in job requirements and employee profiles are emerging, implicitly impacting the reduction on the share of manual work (Müller et al., 2018).

Considering the growing consumer demand and globalization of markets, there is an increasing need for companies to strengthen their competitive advantage (Wang et al., 2016). The implementation of digitization enables better monitoring of processes

Technical University of Iaşi, Faculty of Industrial Design and Business Management, Str. Prof. Dr. Doc. Dimitrie Mangeron, Nr. 29, Iaşi, Romania.

* Corresponding author: constantin.loghin@academic.tuiasi.ro, irina-elena.marin@academic.tuiasi.ro

DOI: 10.1201/9781003518990-7

and hence, improving efficiency, quality, transparency and flexibility (Müller et al., 2018). Moreover, by having smart connected products and access to product data, the manufacturer gains a competitive advantage through the ability and agility to modify and improve product performance (Porter and Heppelmann, 2014). The apparel industry is one of the industries that is characterized by massive labour involvement, a wide variety of product designs, a high volume of production, high market competition and high demands regarding the products' quality (Gereffi and Korzeniewicz, 1994).

The introduction of the latest technological advances (3D printers, automation and robotics and advanced transportation) in garment production will modernize the industry by providing information transparency, autonomous control and sustainable production (Radziwon et al., 2014). This transformation will be essential for the fashion industry because today the industry is facing major problems in terms of "market research, competition monitoring, investing in its own development, original creation and scientific approach to the introduction fashion products to the market" (Colovic and Lamotte, 2014; Parviainen et al., 2017).

The digital transformation of fashion companies offers a competitive advantage as it offers flexibility and agility by quickly picking up new fashion trends and reacting quickly to them, leading to minimized human intervention on processes, increasing productivity and exploiting resources more efficiently (Porter and Heppelmann, 2014; Jayatilake and Withanaarachchi, 2016). For example, cloud computing integration in the fashion industry not only facilitates access to all stages of the manufacturing process anywhere and anytime, but also promotes product customization and clear transparency of the entire purchasing process (Jayatilake and Withanaarachchi, 2016).

The digitalization process began in 2011, with the conceptualization of Industry 4.0. Since the development of the first computer in 1939 and the advent of accessible internet in 1991, the adoption of digital solutions in industrial processes has experienced exponential growth. The evolution of digital technologies is marked by significant dynamism, raising concerns about potential negative social impacts, such as those associated with AI technologies. Prior to 2020, the pace of digital transformation was relatively slow. However, the COVID-19 pandemic triggered the advancement of global digitalization, prompting unprecedented allocations of public resources to support this transition.

The fashion industry, one of the oldest and largest global sectors (Mukherjee, 2015), encompasses firms engaged in designing, producing, and selling apparel, footwear, and accessories. Increasing global competition and volatile demand compel fashion manufacturers to continuously enhance their production processes to ensure sustainable business development (Abd Jelil, 2018). Moreover, the apparel sector has historically been an early adopter of digital solutions, in terms of automation, logistics optimization, and design software, which are now extensively utilized within the production systems. According to McKinsey and Company's *The State of Fashion 2021* report, fashion entrepreneurs identify the COVID-19 pandemic and the resulting economic crisis as their greatest challenges, while viewing digitalization as the primary opportunity for survival. Statistical data from EURATEX reveal that in the first half of 2021, sales of apparel products represented only 88.5% of those recorded during the same period in 2019.

Europe's industrial strategy highlights the continent's transition toward climate neutrality and digital leadership, with expectations that European industries, including textiles and apparel, will contribute significantly to these goals. Fashion companies are urged to prioritize digitalization to avoid exclusion from global supply chains, which are already undergoing their own rapid digital transformation.

Industry 4.0 represents the advancement of digital technologies aimed at modernizing industries, in order to address pressing challenges, such as global competition, environmental pollution, and labor conditions (Wang et al., 2016; Müller et al., 2018). From an ecological perspective, Industry 4.0 seeks to mitigate climate concerns and reduce environmental pollution. Its central goal is minimizing waste, resource consumption, and energy use (Kiel et al., 2017). By facilitating data-driven insights into carbon footprints, Industry 4.0 supports efforts to reduce greenhouse gas emissions. Additionally, it enables the reduction of incorrect deliveries, waiting times, and defective products, contributing to more efficient resource and energy consumption (Kiel et al., 2017).

According to Boston Consulting Group's *Pulse of the Fashion Industry 2017* report for the Copenhagen Summit, the fashion industry is responsible for over 4% of global waste production, amounting to 92 million tons annually, and is considered a major polluter (Pal and Gander, 2018).

2. The 4.0 apparel manufacturing system

A company which operates through a production system as presented in Figure 1, featuring fully digitized and interconnected processes, is an intelligent company (Smart Factory) (Gokalp et al., 2018). In an SFF-type company, the customer provides the marketing department with order details (price, quantity, size distribution, material samples, delivery deadline). If the order is accepted, the information is passed on to the production and procurement departments. After determining the required raw materials, the procurement unit identifies suppliers of materials and accessories and initiates the procurement process. The production planning department develops the production plan and, based on the order information, determines the delivery date using operational analysis data, production capacity, and material delivery dates. After receiving the materials, their information serves as input for sorting and cutting, followed by the manufacturing process (cutting, assembly, finishing, packaging), and finally delivery. Activities within each department are fully digitized, with complete digital interconnection between departments.

Table 1. Industry 4.0 tools available for the apparel manufacturers.

<table>
<tr><td colspan="2">Digital Data Transfer</td></tr>
<tr><td>Available solution</td><td>Manifested as</td></tr>
<tr><td>3D Virtual Garment Modeling</td><td>Simulating garments within the virtual space</td></tr>
<tr><td>Model Examination in a Digital Environment</td><td>Supported by virtual reality technologies, enabling mannequin wear simulation and client communication.</td></tr>
<tr><td>Cutting Layout Transmission</td><td>Sending patterns to cutting systems via a wireless network in a digital environment using cloud infrastructure.</td></tr>
<tr><td>Cybersecurity for Industrial Devices</td><td>Protecting devices against cyber threats while enabling secure, reliable data communication through flexible network infrastructure</td></tr>
<tr><td colspan="2">Maintenance Forecasting</td></tr>
<tr><td colspan="2">Involves statistical analysis techniques to identify scenarios and establish relationships between devices. It is used to track devices and equipment to predict potential failures and implement maintenance strategies, reducing failure rates and improving device utilization.</td></tr>
<tr><td colspan="2">Cyber-Physical Systems</td></tr>
<tr><td colspan="2">Equipped with sensors, actuators, and smart electronic processors connected to the internet, these systems extract necessary information to optimize decisions. They ensure system functionality and capacity by controlling high-data-flow devices in real time, with self-optimization capabilities to anticipate errors and quality issues.</td></tr>
<tr><td colspan="2">Human-Robot Interaction in the Cutting Room</td></tr>
<tr><td colspan="2">Material transport from storage to the cutting room via cyber-physical systems. Fabric spreading and cutting operations are performed using cyber-physical systems and robots, with minimal human interaction.</td></tr>
<tr><td colspan="2">Smart Manufacturing Systems</td></tr>
<tr><td colspan="2">• RFID Tags: Placed on cut products, these tags store information about manufacturing, finishing, and packaging technologies.
• Production Tracking: RFID readers at each station collect real-time production data, guiding products through the next steps.
• Automatic Parameter Adjustment: Technological equipment can self-regulate parameters based on RFID data without human-machine interaction.</td></tr>
<tr><td colspan="2">Robotic Quality Control</td></tr>
<tr><td colspan="2">Quality standards include dimensional accuracy, material quality, and finishing. Industry 4.0 integrates video inspection systems for faster, error-free quality control and real-time production data collection.</td></tr>
<tr><td colspan="2">Packaging with Cyber-Physical Systems</td></tr>
<tr><td colspan="2">Packaging is guided by RFID information, ensuring no human interaction with the systems.</td></tr>
</table>

By analyzing the structure depicted above, we can further analyze the ways in which Industry 4.0 has found its applications within the manufacturing sector (Table 1). The solutions that have been identified and depicted below, not only contribute to the reduction of the negative impact that the manufacturing process has over the environment, but they also provide time optimization, which is a crucial factor within any enterprise.

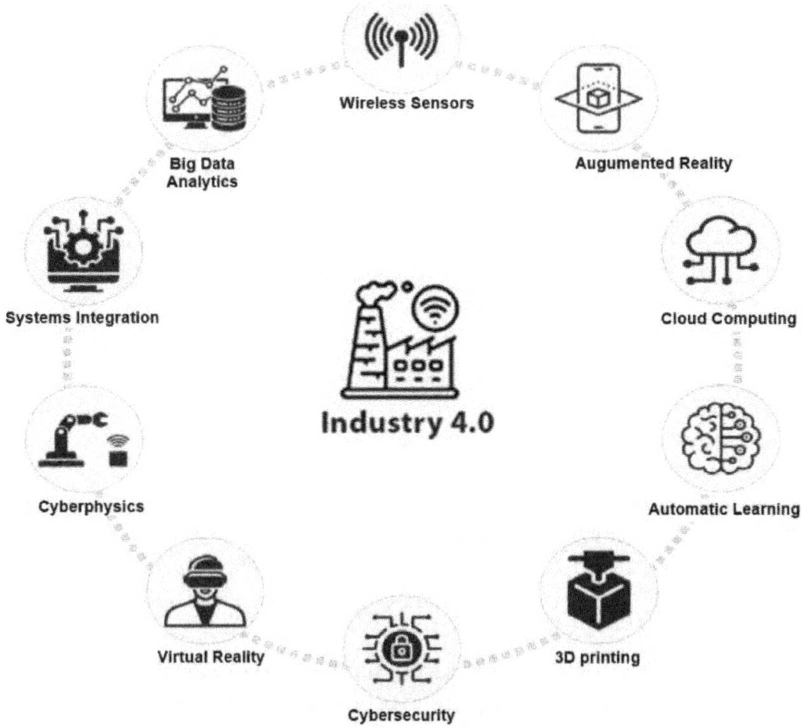

Figure 1. The digital tools available for the manufacturing system (Gokalp et al., 2018).

3. The digital transformation within the manufacturing companies – case study

In the broadest sense, the manufacturing industry involves the physical or chemical transformation of materials, substances, or components into new products. The manufacturing process requires the use of labor, machinery, tools, and energy resources. A comprehensive analysis of a company's digital transformation process must be conducted from the perspective of its interconnection with supply chain elements, the modularization of the manufacturing system, the digital interfacing of production means, product traceability, and the online monitoring of processes.

Between October 2018 and March 2019, an extensive study was conducted to analyze the process of adopting Industry 4.0 principles in Romanian companies operating in the industrial sector (Digitalization in industrial companies in Romania, Perception Study, 2019). The study was carried out by two Romanian consulting firms (Cult Market Research and Academia Industrială) on a sample of 300 companies. The methodology used for data collection involved administering a questionnaire with 57 questions, either online or onsite, to representatives of the selected companies. The main conclusions of the study address aspects related to the perception of digitalization, the digitalization strategy, and the performance of the digitalization process, depending on the industrial profile and level of development.

- **The significance of the digitalization process for companies in Romania:** For approximately one-third of the interviewed companies, digitalization has recently become a priority, indicating that the Fourth Industrial Revolution is still in its early stages in the country. The technologies implemented so far are effective, but Industry 4.0 brings innovations that involve a rapid pace of change, which the Romanian industry should follow to maintain competitiveness. It has been observed that production planning and control are the priority areas for implementing digital solutions. This is expected, as companies have initiated the digitalization process to increase efficiency, gain better control over activities, and reduce manufacturing time.
- **The Pace of Implementing the Digitalization Strategy in Romanian Companies:** Romanian industrial companies are aware of the benefits brought by digitalizing their activities. However, at present, the mass implementation of digitalization across companies is progressing at a very slow pace. The commitment made at the management level is the main driving force behind the implementation of digitalization, with 44% of the surveyed companies receiving constant support for digitalization from management. The success of the digitalization process is guaranteed only if management considers it a priority and follows all the necessary steps according to a well-defined digitalization strategy.
- **Investment planning in digitalization for Romanian manufacturers**: The vast majority of companies (97%) have expressed a desire to invest in digitalization in the near future, even though they do not yet have a strategic plan. This is promising, but should be viewed with caution, as companies have highlighted the lack of a favorable context for digitalization (as of 2019) in all respects. The areas where manufacturing companies intend to make significant investments are planning and monitoring production, which are also considered the priority areas for initiating the digitalization process.
- **The factors influencing the digitization of Romanian companies:** Apparel manufacturing companies have different needs regarding the implementation of the digitization process. However, participants face a shortage of skilled labour for activities requiring digital competencies and experience greater financial difficulties compared to companies in other sectors. They also intend to invest more in the digitization of administrative and decision-making processes compared to companies in other industrial fields.

The *garment prototyping*, as well as *logistical processes on garment lines* are the main processes which hold a high digitalization potential for the fashion companies, and for which software have already been developed.

However, sewing and wet-thermal processing operations for apparel items still contain a low digitalization potential due to numerous influences from textile materials, and the diversity of patterns (Verzea and Luca, 2006).

3.1 Trends in the digital transformation of companies in the fashion industry

The fashion industry is undergoing a paradigm shift in production, as the market trend is towards cost and low environmental impact. The trend of moving production to countries such as Bangladesh, Vietnam and Cambodia with low labor costs, seems to have come to an end as energy and fuel costs are rising. In addition, the high carbon footprint is due to the distance between suppliers, manufacturers and customers. In this context, companies in the fashion industry need to increase product quality, efficiency and ultimately eliminate waste.

At European level, the Commission has identified and published the report Technological trends in the textile industry. After analyzing the websites of fashion companies, the report provides strategic directions for the development of the fashion industry (Figure 2) provide an overview of the use of digital technologies in the digitalization process of the fashion value chain.

Companies that implement and use digital technologies at all stages of the value chain are considered as "smart fashion factories".

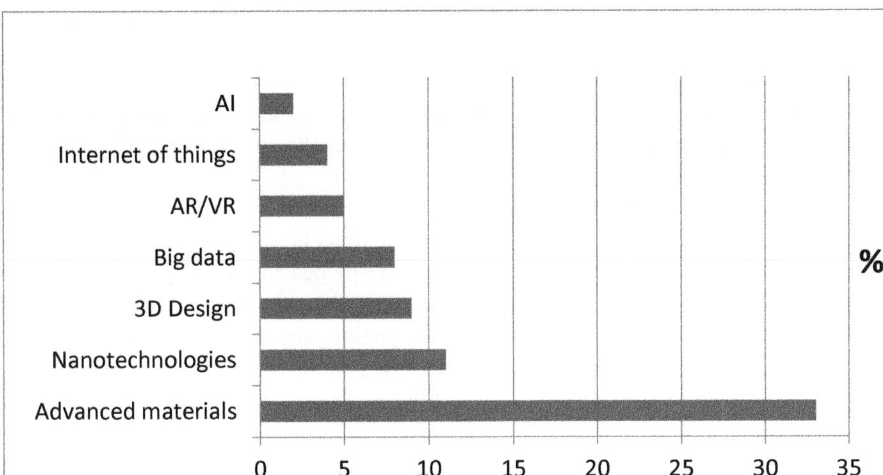

Figure 2. Future strategic directions necessary for fashion industry companies.

4. Design innovation and virtual prototyping: reducing waste within the design process

Textile fabrics are the main resource in the textile manufacturing industry. Due to their versatility, they serve as both finished products and raw materials in the garment manufacturing process. With technological advancements, textiles are available as woven, knitted, or nonwoven structures. Defined by their rigid, elastic, or unconventional nature, these structures are essential in our lives through the products they create. The manufacturing process begins with the cutting operation, where the textile surface is sectioned or cut according to the geometry of the patterns.

Before assembly, the components of the product to be created are cut from the designated textile material(s). Pattern layouts can be created manually or digitally. To increase efficiency and save time, dedicated software has been developed, enabling both manual and automated placement based on predefined algorithms. Two main elements are required for the cutting operation: the textile surface and the corresponding marker, containing patterns arranged according to an integrated algorithm that the program operates on.

The fabric waste generated during the manufacturing processes of a textile article, before it is put into use, is called pre-consumer waste. This waste originates from various phases and operations throughout the entire technological flow. If the fabric scraps resulted from the cutting process are not intended for recycling, they are usually destined for destruction, most often through incineration (Figure 3). Although at first glance, burning may seem like a more efficient method of managing textile waste compared to depositing it in large quantities in open areas, this method generates a series of pollutants harmful to health (Šajn, 2019).

Incineration releases chemical compounds into the atmosphere present in dye formulas used for coloring fabrics, chlorine particles from bleached materials, and plastic particles from fibrous layers or coated textiles. Among the volatile organic compounds emitted during textile waste incineration is methane, exposure to which can cause migraines or dizziness in humans. Not all textile waste gets a second life through recycling, and poor waste management results in adverse environmental effects, whether in solid or gaseous form.

The constructive design of clothing using the Zero Waste (ZW) technique aims to reduce textile material waste by up to 100%. This is achieved by geometrically reshaping traditional sewing patterns so that the pieces interlock and complement one another when laid out on a single fabric unit. Patterns are interconnected, and their placement resembles the arrangement of puzzle pieces. This technique offers a preventive solution to the generation of textile waste by ensuring the full utilization of the textile surface. The defining characteristic of this constructive design method

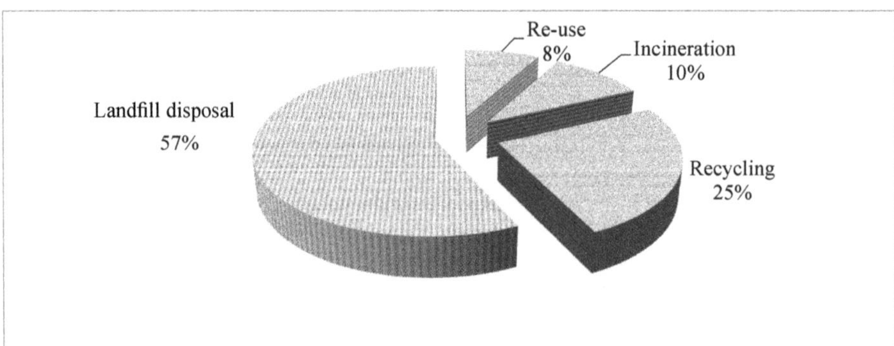

Figure 3. The share of methods used to manage textile fabric waste (Common Objective, 2017. Fashion and Waste: An Uneasy Relationship).

lies in both significantly reducing and preventing waste formation (in this case, material waste). This outcome is the most desirable and favorable option for the textile manufacturing industry (Figure 4).

The hierarchy of waste management options was established and introduced under Directive 75/442/EEC. Following amendments, a new directive, 2006/12/EC, was instituted to clarify, improve, and replace the previous one.

By determining the resources needed for the production of textile supports, it was possible to evaluate the value of a material and the losses resulting from the cutting process. It was established that material waste is inevitable due to the geometry of patterns, which does not allow the entire surface to be used or occupied, resulting in pieces of material that cannot be reused in the future (due to design rules or their small dimensions). Thus, textile material waste management is carried out through four known and commonly used methods in current practice, which differ in their degree of application.

A layplan (also referred to as "marker") is considered efficient if it achieves or exceeds 86% material utilization, also known as layplan's efficiency. This value is automatically calculated and indicated by the CAD systems that are specifically dedicated for layplan making. By using such digital tools, the potentially unused textile material, commonly referred to as "textile losses" within the manufacturing industry, can be easily identified as being the areas where patterns can't be fitted into.

In order to give a representative example of this notion, a comparison has been made between two women's blouses, that showcases, first and foremost, the core differences in terms of marker efficiency and fabric consumption, between a set of classical and ZW patterns (Figure 5). It's important to mention that the markers have been generated using the same fabric width in both cases (140 cm). At first glance, it is obvious that the potential fabric losses (Figure 5 A: hatched areas) are significantly wide. As aforementioned, the key pieces of information are displayed by the software (Gemini Nest Expert) on the bottom ribbon of the layplan. These values indicate, not only that the ZW patterns lead to the reduction of the potential fabric waste (given by the 99.79% layplan efficiency), but they only ensure a reduction of the

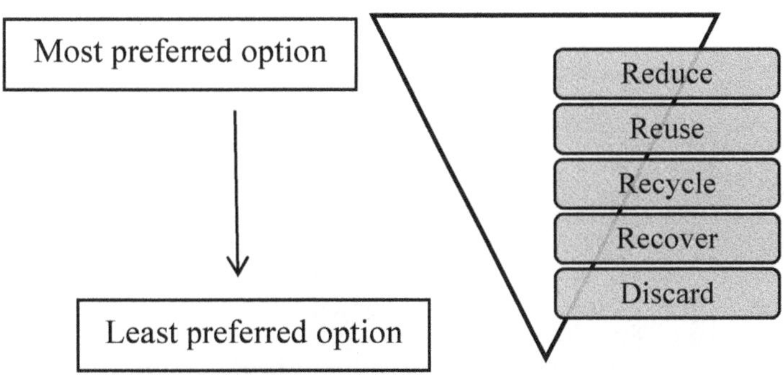

Figure 4. The hierarchy of waste management (Directive 2006/12/EC Of The European Parliament And Of The Council of 5 April 2006 on waste).

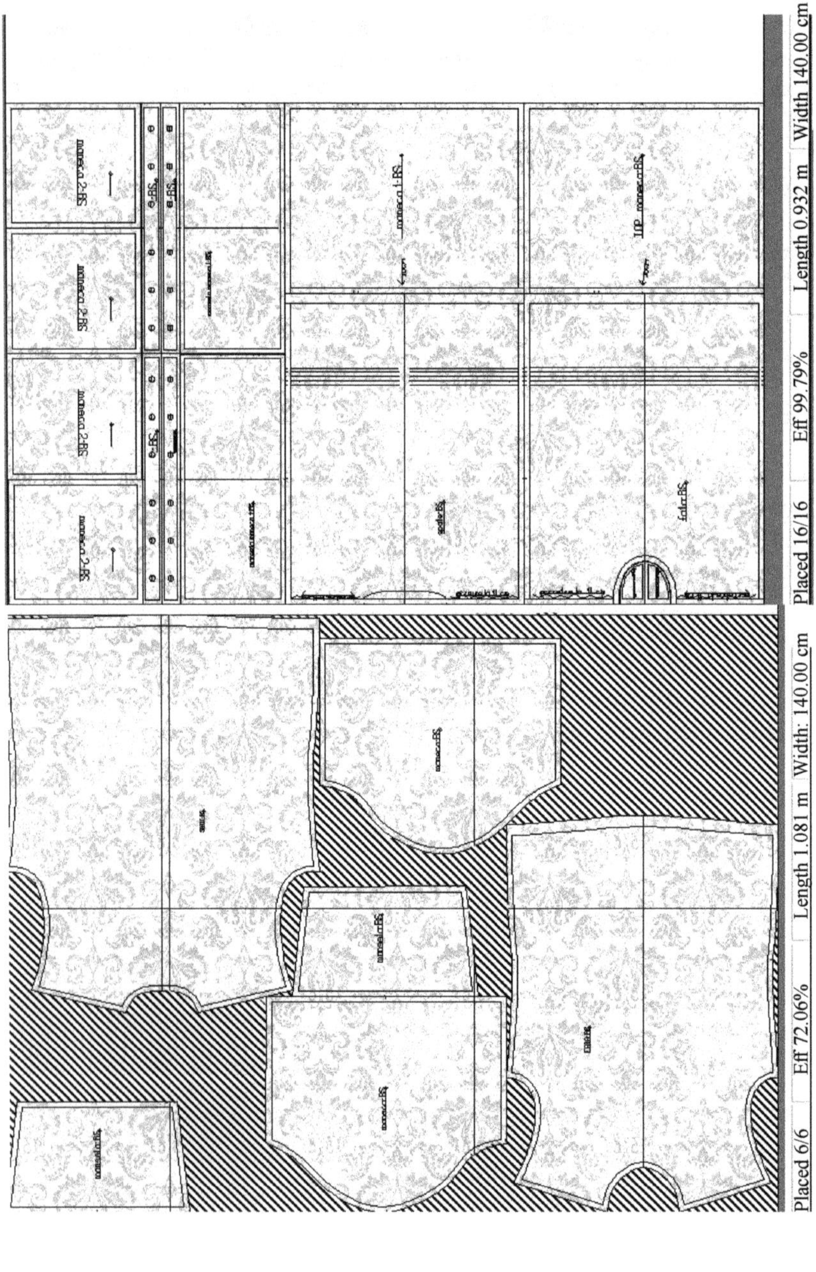

Figure 5. Marker efficiency and fabric rating comparison between a set of classical patterns (left) and a set of ZW patterns (right).

fabric consumption by 0.149 meters – for one unit. That would translate itself into 1.49 meters for a production order composed of 10 units.

At the core of any sustainable transformation in fashion lies the design phase, which is responsible for determining material usage and production strategies. Traditional design practices often lead to substantial waste, both in terms of unused fabrics and excessive samples. The digital transformation of this phase has the potential to mitigate such waste.

3D design and virtual prototyping have revolutionized the fashion sector, by allowing designers to create and modify designs digitally before physical production begins. This eliminates the need for multiple physical samples, which often result in surplus materials and discarded prototypes. It is well known that by following the traditional practices of sampling, a significant amount of fabric scraps are generated. The textile waste not only brings a negative impact over the environment, as previously stated, but they also play a significant role on the apparel item's carbon footprint.

Tools like CLO 3D and Style 3D enable designers to simulate fabric behavior, test color combinations, and preview garments without committing to physical sampling, significantly reducing material waste in the early stages. By following the dedicated digital assets, garments can be created within the virtual space, known as the so-called "digital-twin" attributed to the corresponding apparel item (Marin et al., 2024).

These digital tools provide a practical and efficient solution for optimizing time, energy, and processes involved in sampling. Simultaneously, they serve as a powerful asset in the co-creation process, enabling the exploration of different shapes and stylistic variations without relying on traditional physical prototypes. Thus, this digital approach prevents the usage of any fabrics and trims, leading to the minimization of fabric waste.

Any textile material unit, considered unwanted and useless, is transformed into waste. Since the waste resulting from the cutting process is a major contributor to pollution caused by the textile industry, the need for implementing techniques to reduce the production of such waste is increasingly emphasized. The separate collection of textile waste is a measure that must be implemented by 2025, according to the European Parliament. However, the challenge lies in changing, understanding, and replacing or improving the traditional approach to constructive design with one that addresses the issue of material scraps in advance (Marin, 2024). When it comes to the fabric waste that is typically generated as a result of the pattern cutting, it's worth mentioning that an effective option consists of using the jigsaw layplan structure, widely known to contain *zero waste patterns*.

The design process of the garment includes the pattern drafting procedure. This aspect is mostly accurate, considering the main factors that must be taken into account, from the very start of the pattern drafting process: the fabric's width and the fabric's properties (such as artwork or structure). The later elements are utterly crucial and considered to be the starting points on which the drafting process is gradually built further (Marin, 2024). The nature of those elements can be viewed as the "restrictions" or "limitations" that set the rules responsible for the pattern placement (when print continuity is mandatory), pattern orientation (considering the fabric's structure) and the layplan's width (given by the fabric's width).

Given their characteristic geometrized shape, which tells them apart from their classical counterparts, the ZW patterns are prone to lead to garment items with an inherited different or unusual fit. With the appropriate know-how and by applying the technological methods available, the apparel generated by using the ZW can be shaped and modeled in a significant manner, and thus, the exaggeratedly geometric shape of the ZW patterns shouldn't raise any concerns in terms of the body fitting.

The drafting process can be approached via a number of methods, as long as the aforementioned parameters (the fabric's width and fabric properties) are taken into account. By using the CAM systems and other digital tools available, the ZW patterns can be obtained by shaping a set of already existing classic patterns, and stretching their geometry to the extremes (Figure 6), aiming to generate the puzzle-looking layplan. In this particular case, the study conducted has led to the possibility of designing a modular apparel item, characterized by two sets of detachable sleeves, both in different lengths (Figure 7). Thus, not only that the design process has endured the attainment of a ZW apparel, but the item obtained was characterized by versatility, resulting in a number of ways through which it can be worn. Another important aspect is represented by the fabric's print direction. This attribute represents onto itself a restriction when it comes to the direction in which the patterns will be placed.

This showcases the efficiency of the digital tools, that have been used along the process for drafting the ZW patterns and creating the jigsaw layplan, as well as for simulating the garment item in the virtual space. Thus, the output consists in a series of illustrations, showing the versatility of the garment, and that can further be used to promote the product on specific platforms.

Nevertheless, the virtual space in which the garment has been simulated offers the opportunity to obtain the visual end-result of the size range (Figure 8) for the garment that's being tested. This is also possible by adjusting the avatar's body measurements in according to any given size table. This aspect is made possible due to the shaping techniques that have been used within the garment, consisting of stitches made with elastic thread. Depending on the avatar's body measurements, the gathering areas can be increased or decreased during the manufacturing process.

Digital co-creation technologies significantly shorten the time required for designing products or services. The product design process typically involves creative and design stages using CAD tools, where the manufacturing process must be considered in advance. This may include the need to design specific tools, such as those for spatial shaping processes. Digital tools for participatory design enable remote collaboration, allowing multiple people to review or modify a model. However, these tools are primarily focused on model visualization and lack the capability to fully represent product use or real-time exploitation during the review process.

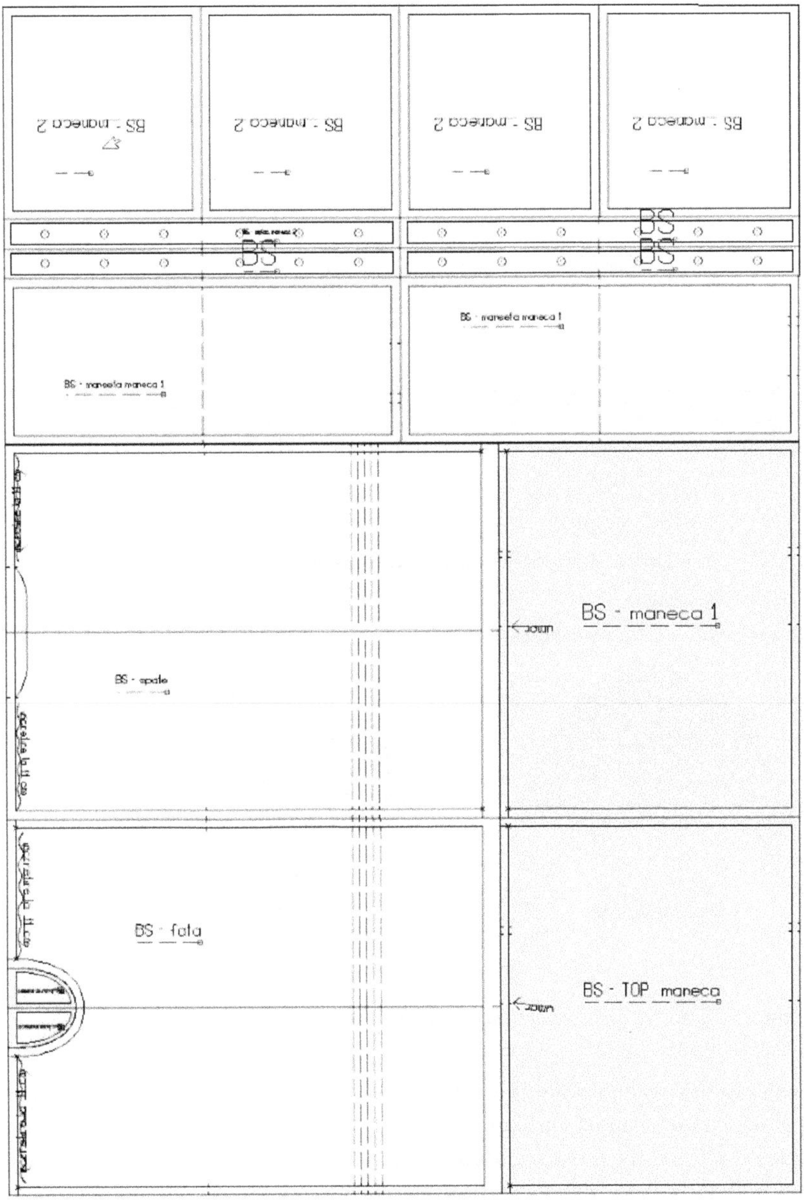

Figure 6. The geometrical method of drafting ZW patterns.

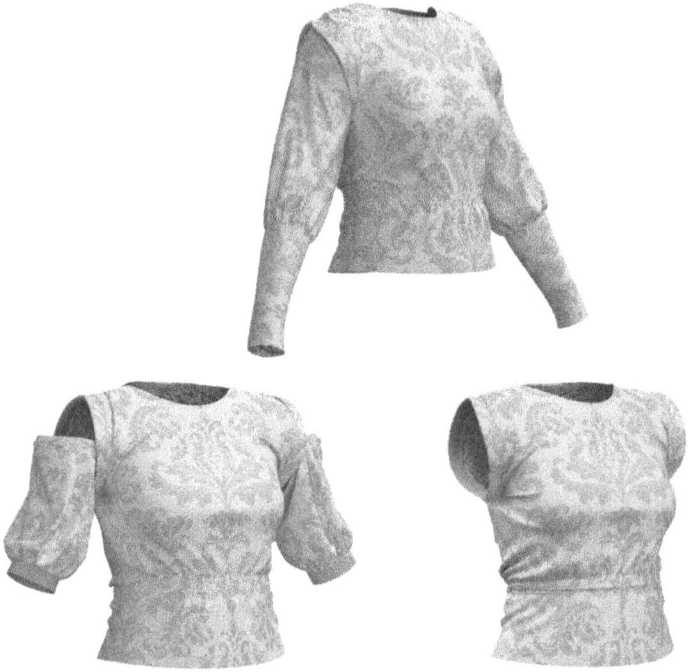

Figure 7. The modular apparel item's virtual prototype.

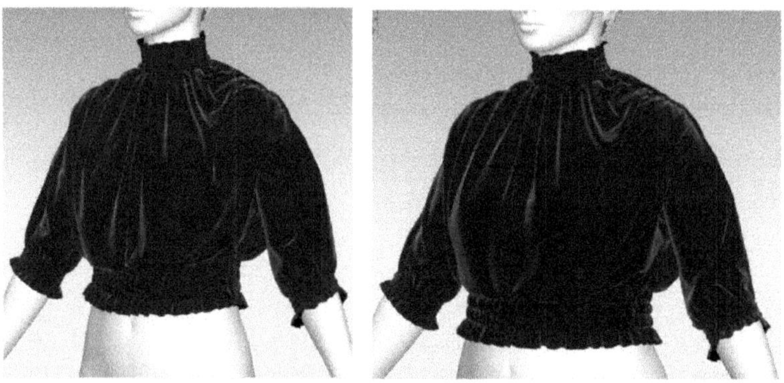

Figure 8. Adapting the garment item on extreme sizes: from S (left) to XL (right).

To accelerate design iterations, this capability is crucial for designers. The process and the product designer must collaborate closely to ensure that the conceptualized product can be efficiently manufactured. Currently, digital tools do not facilitate swift collaboration between these two professionals, hindering the iterative process of design and product development. Moreover, integrating co-creation into the broader digital product development chain is vital. To align with emerging production trends that cater to specific consumer demands, these tools must be customized for different stakeholders (consumers, designers, manufacturers) to create a seamless digital workflow across all participants. For instance, consumer preferences can be directly translated into actionable work instructions or production line adjustments (Loghin et al., 2023).

The harmful effects on the environment generated by the cumulative actions within a product's value chain have necessitated the analysis of all pollution sources, allowing for the conclusion and definition of methods to reduce the impact a product has on the environment throughout its entire lifecycle. The term "Life Cycle Assessment," also referred to as "Life Cycle Analysis," focuses on the methodological approach to environmental management, identifying the extent of a product's environmental impact from the design phase to its disposal. A case study has been conducted, in order to illustrate and quantify the difference that the ZW patterns have upon the garment item's carbon footprint. According to the data obtained, CO_2 emissions were reduced by 1.6% due to the use of a smaller amount of material, for the production of the ZW item (Marin et al., 2023).

5. Conclusions

The digital transformation offers immense potential for a more sustainable, zero-waste fashion industry, but several challenges must be addressed:

- **Technology Costs and Accessibility:** Advanced technologies can be expensive, particularly for smaller fashion brands. Investment in these technologies may be out of reach for many, limiting the ability of small-scale producers to participate in the digital transformation.
- **Consumer Awareness and Education:** While digital tools can promote sustainability, they require an educated consumer base to truly have an impact. Brands must invest in educating consumers about the benefits of zero-waste, sustainable fashion, and the role that digital tools play.
- **Integration with Traditional Practices:** Many legacy manufacturers still rely on traditional methods. The transition to digital, sustainable systems requires retraining and a shift in mindset across the entire fashion supply chain.

Digital transformation is rapidly changing the fashion industry, enabling a shift toward sustainability and a zero-waste model. From design innovation, AI-powered production, and circular systems, to block chain-based transparency, digital tools are helping brands reduce waste, optimize resources, and improve overall sustainability. However, achieving a truly sustainable zero-waste industry will require continued investment, collaboration across the supply chain, and consumer engagement to drive systemic change. As these technologies become more accessible and widespread, the future of fashion looks more circular, sustainable, and resource-efficient.

In conclusion, while digital transformation is not a panacea for all the sustainability challenges faced by the fashion industry, it offers a promising pathway towards a more resource-efficient and zero-waste fashion economy. The continued integration of advanced technologies will not only streamline production but will also encourage sustainable consumption and waste management practices, ultimately enabling the fashion industry to reduce its environmental footprint and foster a circular, regenerative economy.

References

Abd Jelil, R. (2018). Review of artificial intelligence applications in garment manufacturing, Artificial Intelligence for fashion industry in the big data era. doi.org/10.1007/978-981-13-0080-6_6.

Academia Industrială - Digitalizarea în companiile industriale din România. Studiu de percepție, 2019.

Bick, R., Halsey, E. and Ekenga, C. C. (2018). The global environmental injustice of fast fashion. Environmental Health 17: 92. doi:10.1186/s12940-018-0433-7.

Colovic, A. and Lamotte, O. (2014). The role of formal industry clusters in the internationalization of new ventures. European Business Review. doi.org/10.1108/EBR-02-2013-0027.

Common Objective. (2017). Fashion and Waste: An Uneasy Relationship.

DIRECTIVE 2006/12/EC OF THE EUROPEAN PARLIAMENT AND OF THE COUNCIL of 5 April 2006 on waste.

Gereffi, G. and Korzeniewicz, M. (1994), Commodity Chains and Global Capitalism, Economics, Praeger Publishers, 88 Post Road West, Westport. CT 06881, ISBN: 0-275-94573-1.

Gökalp, E., Gökalp, M. O. and Eren, P. E. (2018). Industry 4.0 revolution in clothing and apparel factories: Apparel 4.0, Industry. Industry 4.0 from the MIS Perspective. doi: 10.3726/b15120/21.

Jayatilake, H. and Withanaarachchi, A. S. (2016). Industry 4.0 in the apparel-manufacturing sector: Opportunities for Sri Lanka. 1st Interdisciplinary Conference of Management Researchers.

Kiel, D., Müller, J. M., Arnold, C. and Voigt, K. I. (2017). Sustainable industrial value creation: Benefits and challenges of industry 4.0. International Journal of Innovation Management. doi.org/10.1142/S1363919617400151.

Loghin, E. C., Vîlcu, A. and Verzea, I. (2023). Particularities of the digital transformation of the manufacturing industries. Buletinul Institutului Politehnic din Iaşi 69(73): 1.

Marin, I. E., Bocancea, V. and Loghin, M. C. (2023). The influence of Zero Waste sewing patterns upon the apparel's CO_2 footprint. European Journal Of Materials Science and Engineering 8(2). doi: 10.36868/ejmse.2023.08.02.108.

Marin, I. E., Bocancea, V. and Loghin, M. C. (2024). Developing a zero-waste pattern drafting method suitable for mass production. Industria Textilă Journal 75(2). DOI: 10.35530/IT.075.02.202344.

Mukherjee, S. (2015). Environmental and social impact of fashion: towards an eco-friendly, ethical fashion. International Journal of Interdisciplinary and Multidisciplinary Studies (IJIMS) 2(3): 22–35.

Müller, J. M., Kiel, D. and Voigt, K. I. (2018). What drives the implementation of Industry 4.0? The role of opportunities and challenges in the context of sustainability. Sustainability. doi.org/10.3390/su10010247.

Newman, S. T. (2017). Intelligent manufacturing in the context of Industry 4.0: A review. Engineering 3: 616–630. https://doi.org/10.1016/J.ENG.2017.05.015.

Pal, R. and Gander, J. (2018). Modelling environmental value: An examination of sustainable business models within the fashion industry. Journal of Cleaner Production. doi.org/10.1016/j.jclepro.2018.02.001.

Parviainen, P., Tihinen, M., Kääriäinen, J. and Teppola, S. (2017). Tackling the digitalization challenge: how to benefit from digitalization in practice. International Journal of Information Systems and Project Management vol. 5, no. 1.

Porter, M. E. and Heppelmann, J. E. (2014). How smart, connected products are transforming competition. Harvard Business Review.

Radziwon, A., Bilberg, A., Bogers, M. and Madsen, E. S. (2014). The smart factory: exploring adaptive and flexible manufacturing solutions. Procedia Engineering. doi.org/10.1016/ j.proeng.2014.03.108.

Šajn, N. (2019). Environmental Impact of the Textile and Clothing Industry: What Consumers Need to Know. EPRS: European Parliamentary Research Service.

Verzea, I. and Luca G. P. (2006). Managementul logisticii industriale şi comerciale, Ed. Peformantica Iasi.

Wang, S., Wan, J., Zhang, D., Li, D. and Zhang, C. (2016). Towards smart factory for industry 4.0: a self-organized multi-agent system with big data-based feedback and coordination. Computer Networks. doi.org/10.1016/j.comnet.2015.12.017.

Chapter 8

Green Energy Approaches in Organizations

Alexandra Coroian,[1,*] *Muddassar Sarfraz,*[2] *Timea Cisma,*[1] *Mihai Ardelean*[1] and *și Neta Saptebani*[1]

1. Introduction

Green energy, also known as renewable energy, is energy derived from sources such as solar, biomass, geothermal energy, water and wind. Unlike traditional fossil fuels, these sources of energy produce very little greenhouse gases which is essential in the global focus on environmental care (Rahman et al., 2022). Green energy, by definition is more sustainable, secures our energy security for future generations and prevents adverse climate changes. Ethically and strategically, green energy can be integrated into conventional business processes (Elkhatat and Al-Muhtaseb, 2024). This indicates commitment to the goals for global sustainability and is becoming more important as a source of innovation, economic efficiency as well as organizational resilience.

This direct response to environmental issues like global warming, biodiversity loss and waste is of an increasingly greater focus on sustainability (Peduzzi, 2019). Following sustainable practices have since been underlined by industries, regulators and international agreements, such as the Paris Agreement (Ourbak and Magnan, 2018). Due to this scenario, organizations are in a unique position to serve as change agents. Stakeholders, such as consumers, investors and employees, are applying pressure on them to operate in a way that minimises their environmental impact while

[1] Politehnica University of Timisoara, Faculty of Management in Production and Transportation, 14 Remus Street, Timisoara, Romania.
[2] Faculty of Economics, Administrative and Social Sciences, Nisantasi University, Istanbul, Turkey.
Email: muddassar.sarfraz@gmail.com
* Corresponding author: alexandra.coroian@student.upt.ro

DOI: 10.1201/9781003518990-8

producing long term value. The implication is that sustainability is not an expense anymore but an opportunity to enhance competitiveness and reputation. All of which justifies why renewable energy is a key part of this shift: it offers a practical means to reduce emissions, conserve resources and build resistance to volatile energy markets (Hassan et al., 2024).

Promoting the transition to green energy involves organizations, too. Their activities have far reaching implications as major energy users, as well as contributors to worldwide supply chains. They foster renewable energy facilities like solar panels and windmills, and even smaller scale options, to significantly reduce their dependence on fossil fuels, like solar power systems, small power efficient building designs (Aien and Mahdavi, 2020). Additionally, their leadership tends to cover more ground internally to make a name for themselves in the industry, but more importantly, to spark change. Green energy organizations are the model for other businesses to follow: sustainable practices are feasible even at the expense of profit and success (Song et al., 2019).

This study analyzes how renewable energy serves multiple functions in organizations, both financial, environmental, and societal. It then discusses many renewable energy sources, efficient means of implementation and the obstacles faced by organizations when switching over. Further developments to renewable energy consumption influenced by current state are also discussed, including advancing technology and regulatory advances. This study can contribute to further research to create a more sustainable, fair and resilient organizational future, with the introduction of green energy by making certain recommendations in this regard.

2. Importance of green energy in organizations

The difference between green energy practices among companies has a very decisive effect on both global environmental matters and those of economic implications particularly where green energy practices are promoted (Al-Shetwi, 2022). By switching to green energy sources, organizations can significantly reduce their carbon footprints, gain financial benefits, enhance their brand, and ensure regulatory compliance (ZOU et al., 2021). Green energy has been identified to be important for assisting long term success of improvement in corporate process.

2.1 Environmental impact

The greatest advantage of using green energy is that it lowers carbon footprints and limits climate change. The main sources of carbon dioxide emissions are traditional energy sources, mostly fossil fuels, that impose global warming and other environmental problems (ZOU et al., 2021). When organizations move toward green energy such as solar, wind, or geothermal, they can minimize their emissions. For instance, taking away CO_2 emissions from coal fired electricity is not only a zero CO_2 end product, but also a zero N_2O and SO_2 emission one (Kabeyi and Olanrewaju, 2022).

In addition, green energy use relates to other broader environmental ambitions, including biodiversity conservation and natural ecosystem protection. Abundant and renewable resources like sunlight and wind is what renewable energy systems commonly use (Aeknarajindawat, 2020), to avoid the need for environmentally damaging extractive businesses. The shift supports global climate initiatives, especially the Paris Agreement, in recognizing its overarching objective of keeping global warming below 2°C (Ourbak and Magnan, 2018).

2.2 Economic benefits

Green energy is beneficial for the environment but has even greater economic benefits to businesses. Cost reduction is one of the most appealing advantages. With the increase in technology and a wider use, renewable energy technologies that were too expensive previously have become more feasible (Vanegas Cantarero, 2020). One example, is that solar energy has become affordable for businesses due to the value of photovoltaic systems having dropped in the last decade (Choudhary and Srivastava, 2019). Also, wind energy has low running expenses and supplies which is another economically feasible method of reducing energy costs (Khan et al., 2018).

Reduction in costs is also associated with energy efficiency. Lowering energy use by introducing LED lighting, smart building systems and improved HVAC technologies can save organizations more than half of their energy use (Gökgöz, 2018). Such reduced costs not only raise profitability but can also act like a safety net against the extreme variations of fossil fuel expense. Not to mention embracing green energy may result in other earning sources like carbon credits or subsidies offered by governments to organizations carrying actions of renewable energy.

2.3 Reputation and branding

By adopting green energy, an organization increases significantly their reputation and branding. Today consumers, shareholders, and staff people are looking for businesses that are committed to being green. A friendly competition can also be developed among the businesses adopting green energy that have the opportunity to present themselves as pioneers that show that they are green champions, giving them a competitive edge in the industry.

For instance, the brand image of organizations such as Google and Apple improved greatly by running their operations with 100% renewable energy (Pakulska and Poniatowska-Jaksch, 2022). Not only do these measures appeal to environmentally conscious consumers, but also they draw in the best people to work for companies with strong sustainability ideals. Green energy also can provide an opportunity to strengthen the relationship with its stakeholders, as an example, investors who are increasingly taking into account the ESG factors when making an investment decision (Ziolo et al., 2019).

It also helps customer loyalty and brand equity by connecting with the principles of stakeholders. Often times, believing in a sustainable product leads to long running results, i.e., customers who become sales advocates. Investing in renewable energy shows that the organization is responsible, takes an active role in global efforts to have a sustainable future, and saves money.

2.4 Regulatory compliance

Nowadays, authorities and international agencies are progressively enforcing stricter boundaries to avoid climate change and promote renewable power consumption (*United Nations Climate Change*, n.d.). Failure to follow these standards is bad news to organizations, which can become targets of legal or financial consequences. For instance, carbon pricing and emissions reduction programs affect companies whose greenhouse gas (GHG) emissions cannot be effortlessly reduced. Green energy allows companies to stay ahead of the curve regarding regulatory standards and operate within compliance risks (Ibrahim Adedeji Adeniran et al., 2024).

Moreover, it often pays to switch to green energy because meeting such standards can often also earn an organization rewards, subsidies or tax breaks. For example, many governments grant tax credits for those who install solar panels or offers incentives for buying energy efficient equipment (Villca-Pozo and Gonzales-Bustos, 2019). These will certainly compensate the initial investment expenses not solely but it also guarantees long term savings and compliance.

Similarly, renewable energy usage also lays the foundation for future regulatory changes for organizations. Businesses who have made the commitment to green investing will be positioned to accommodate new needs in a fast changing global business environment (Ourbak and Magnan, 2018).

By using green energy, organizations can get around the environmental problems at hand while advancing economically, increasing their brand reputation, or meeting regulatory needs. These advantages demonstrate the critical significance of green energy in developing a sustainable and resilient administrative future.

3. Types of green energy adopted by organizations

Organizations are rapidly realizing the possibility of renewable energies to transform their energy consumption. Solar, wind, geothermal, biomass, and hydropower are some of the most promising renewable energy sources. Each has distinct benefits and uses, allowing companies to implement solutions that are in line with their operational demands and sustainability goals.

3.1 Solar energy

Solar energy uses sunlight to produce electricity or heat using solar panels and PV, or photovoltaic, devices (Prăvălie et al., 2022). These technologies are extremely versatile, with uses that vary from supplying office buildings to managing large-scale industrial activities. Organizations frequently set up solar panels on buildings or unused land to create electricity on-site, decreasing their reliance on fossil fuels and reducing energy expenses (Choudhary and Srivastava, 2019).

One of the most important benefits of solar energy is its ability to grow. Small businesses can install rooftop systems, but larger companies can build solar farms. Furthermore, developments in photovoltaic technology have increased efficiency and decreased the price of installation, rendering solar energy more affordable. Battery storage systems increase their value by conserving unused electricity for use during low-light periods, ensuring an uninterrupted power supply (Worku, 2022).

Case Study: IKEA

IKEA exemplifies a corporation that has successfully used solar energy. The multinational furnishings manufacturer installed solar panels on the rooftops of several of its stores and warehouses, resulting in significant energy savings and a lower carbon footprint. IKEA has about 1.5 million solar panels in the United States by itself, which generate around forty percent of the energy required for company operations. This determination to use renewable energy is consistent with IKEA's overall sustainability policy, which seeks to make the company climate-positive by 2030 (Mahalakshmi et al., 2024).

3.2 Wind energy

Wind energy, generated by windmills that transform wind movement into power, is another popular renewable energy source. Wind power plants can be built onshore or offshore, depending on geography and operational requirements (Rahman et al., 2022). Businesses that operate in windy areas can particularly profit from this technology, which provides an excellent production of energy compared to the installation cost.

Wind energy efficiency has improved dramatically in recent years. Modern turbines are intended to function at slower speeds of the wind, making wind energy feasible in previously unfavorable areas. Furthermore, wind energy systems have minimal running costs and a relatively quick payback period, which increases their economic feasibility.

Nevertheless, the initial investment in wind turbines can be significant, and their efficiency is dependent on regular wind supply. To address these problems, companies frequently engage with green energy suppliers or engage in power purchase contracts (PPAs) (Stanitsas and Kirytopoulos, 2023), ensuring predictable energy costs while promoting the construction of wind energy infrastructure.

Example: Google

Google has made significant investments in wind energy, ranking among the world's biggest company purchasers of renewable energy. Wind farms provide much of the company's electricity, greatly reducing its dependency on fossil fuels (Lacal-Arántegui, 2019). This program is part of Google's objective to run totally on carbon-free power by 2030.

3.3 Geothermal energy

Geothermal energy uses heat from the Earth's interior to heat, cool, or generate power. It is a very dependable and renewable energy source that provides constant access, whatever the weather conditions (Rahman et al., 2022). Organizations often employ geothermal energy sources for both cooling and heating buildings, utilizing ground-source thermal pumps to transmit heat from the Earth to the building.

Geothermal energy has a huge advantage in terms of efficiency. Heat pumps utilize substantially less electricity than both traditional heating and cooling, systems, resulting in cheaper energy bills and lower greenhouse gas emissions (Paulillo,

2021). Furthermore, geothermal systems have a lengthy lifespan, which increases their cost-effectiveness over time.

Industry Example: Boise, Idaho

The city of Boise exemplifies the successful application of geothermal power in an organizational setting. Boise has one of the most extensive district heating networks in the United States, with geothermal heating serving more than 90 structures, including government agencies and commercial spaces. This system has drastically decreased energy expenditures and emissions, demonstrating geothermal energy's capacity for large-scale applications (Beardsmore, n.d.).

3.4 Biomass and bioenergy

Biomass and bioenergy use organic materials like farm waste, wood, and animal wastes to generate energy, heat, or biofuels (Rahman et al., 2022). This green energy source is especially useful for businesses in the forestry, agriculture, and food service sectors, where bioenergy byproducts are abundant.

The fundamental benefit of renewable energy from biomass is that it can be used to manage trash as well as provide energy (Janiszewska and Ossowska, 2022). Organizations can save money and the environment by transforming organic waste into energy. Furthermore, bioenergy sources can be constructed to function at different scales, making them useful for both small and large industries.

Example: Coca-Cola

Coca-Cola has incorporated biomass energy into their operations at several plants around the world. For example, the organization uses biomass boilers in its plants to produce heat from agricultural waste like sugarcane leftovers. This initiative has lessened Coca-Cola's reliance on non-renewable energy sources and its greenhouse gas emissions (Adeniyi et al., 2022).

3.5 Hydropower

Hydropower, the generation of electricity from flowing water, is one of the oldest and most reliable green energy sources (Rahman et al., 2022). Small scale hydropower systems offer practical alternatives to large scale hydropower projects. Power can be generated on site through these systems, which can be built on streams, rivers or reservoirs.

Hydropower systems are of particular use to groups in rural or isolated areas. They are reliable and provide a great source of green energy while needing low maintenance (Costea et al., 2021). Additionally, these systems could work independently or help other types of green energy to boost energy reliability.

Example: Micro-Hydro in Nepal

Local organisations and people in Nepal have been given access to reliable electricity via micro-hydropower projects. Power to small scale companies, schools and healthcare institutions allow them to remove diesel generators and contribute to economic development. This strategy shows the flexibility of hydropower in accommodating institutionalized and community needs (Poudel et al., 2021).

Green energy sources like solar, wind, geothermal, biomass, hydropower, etc. can save organizations lots of money and reduce their environmental impact. Different advantages are associated with each energy source and this, combined, means that companies can adapt their green energy strategy to their specific needs and resources. The development of a green and durable future is made possible by expanding energy options.

4. Challenges and barriers

Although the existence of renewable energy provides numerous benefits, companies often face overriding obstacles while trying to introduce sustainable energy systems. Obstacles include high expenses, technological and infrastructure limit, reluctance to change, lack of knowledge and regulatory uncertainty (Lu et al., 2020). It is important to address these challenges so as to ensure the effective realization of the green energy plans.

4.1 High initial costs and budget constraints

The high starting costs of green energy systems is one of the most serious constraints to green energy adoption. Some capital expenses to set up windmills, solar panels or geothermal systems are not possible for small or medium sized companies (SMEs) with limited resources. Organizations may have difficulties allocating sufficient resources during the early stages while balancing the costs with sustained benefits (Moorthy et al., 2019). On top of that, it doesn't always get funded, especially in areas where a government program is lacking.

4.2 Technological limitations and infrastructure issues

The implementation of green energy systems is usually effective, but relies on complex technologies and appropriate infrastructure. As an example, wind as well as solar can add to complexity and costs as well as the energy storage needed to support a steady power supply (Guelpa et al., 2019). The adoption process may be made worse by poor infrastructure like out of date electrical grids. Since renewable energy resources may be harder to procure, or dependable, organizations working in rural areas or less developed places may run into additional challenges.

4.3 Resistance to change and lack of organizational expertise

Another major obstacle is internal resistance to change in an organization. Staff and administrators can be hesitant to believe that green energy sources will work or be reliable if, for instance, they have no experience with the technology. Resistance might intensify if it is the first time a company is managing and executing green energy initiatives (Prehoda et al., 2019). For example, if companies do not have in-house sustainability experts or technical specialists, it can be challenging to coordinate green energy adoption with the complexity inherent in doing so, leading companies to get behind schedule or implement inefficiently.

4.4 Regulatory and policy uncertainties

Inconsistent and unpredictable environmental regulation for green energy often makes it difficult for organizations to plan long term investment. Green energy policies and incentives are location dependent and will change as political priorities change (Hu et al., 2023). Improvements in financial feasibility are also sensitive to changes in incentives or tax credits, for example. Additionally, obtaining clearance to compliance standards can be time consuming, severely long and hard, even in countries where the regulatory landscape is unplanned.

To overcome these problems, a comprehensive solution is needed, including borrowing money, developing organizational competence, and proactively getting the regulation brought to a stable state. Dealing with these constraints will see organizations get the maximum value out of the renewable energy and contribute to sustainable future.

5. Methodology

To begin data collecting process, we applied a structured questionnaire as the key instrument for gathering statistical data. A quantitative method was used, with data collected using a standardized questionnaire targeting different organizations. The questionnaire evaluated company features (e.g., size, sector), actual acceptance of renewable energy approaches, and perceived barriers in implementing green energy.

Targeted sampling was applied to find businesses from multiple sectors, like manufacturing, service sectors, and technology. A total of twenty organizations were examined. The questionnaire data has been evaluated using PSPP software.

The chi-square analysis found significant relationships between organizational factors and the uptake of renewable energy practices. Major results include: larger businesses have more financial as well as technical assets, facilitating green energy adoption; the service sector, with lesser facilities constraints, have higher adoption rates than manufacturing; and financial constraints remain a significant barrier, emphasizing the demand for specific grants and benefits.

55% of the analyzed organizations are already using green energy for at least a part of their business. This is mostly encountered in large business, followed closely by small organizations, trying to be more sustainable.

Renewable energy such as solar was the most popular green energy solution with a total of 25%, mainly in medium and large companies, which was followed by hydroelectricity (20%) used mostly in small organizations, and geothermal energy (10%) (Figure 1, Figure 2, Figure 3).

Organization size × Adoption of green practices

			Adoption of green practices		
			Yes	No	Total
Organization size	Small	Count	4	3	7
		Row %	57,1%	42,9%	100,0%
		Column %	36,4%	33,3%	35,0%
		Total %	20,0%	15,0%	35,0%
	Medium	Count	2	4	6
		Row %	33,3%	66,7%	100,0%
		Column %	18,2%	44,4%	30,0%
		Total %	10,0%	20,0%	30,0%
	Large	Count	5	2	7
		Row %	71,4%	28,6%	100,0%
		Column %	45,5%	22,2%	35,0%
		Total %	25,0%	10,0%	35,0%
Total		Count	11	9	20
		Row %	55,0%	45,0%	100,0%
		Column %	100,0%	100,0%	100,0%
		Total %	55,0%	45,0%	100,0%

Figure 1. Adoption of green practices. Source: Authors' own research.

Organization size × Types of green energy practices adopted

			Types of green energy practices adopted					
			Solar	Wind	Hydropower	Geothermal	None	Total
Organization size	Small	Count	1	0	2	1	3	7
		Row %	14,3%	,0%	28,6%	14,3%	42,9%	100,0%
		Column %	20,0%	,0%	50,0%	50,0%	37,5%	35,0%
		Total %	5,0%	,0%	10,0%	5,0%	15,0%	35,0%
	Medium	Count	2	0	0	1	3	6
		Row %	33,3%	,0%	,0%	16,7%	50,0%	100,0%
		Column %	40,0%	,0%	,0%	50,0%	37,5%	30,0%
		Total %	10,0%	,0%	,0%	5,0%	15,0%	30,0%
	Large	Count	2	1	2	0	2	7
		Row %	28,6%	14,3%	28,6%	,0%	28,6%	100,0%
		Column %	40,0%	100,0%	50,0%	,0%	25,0%	35,0%
		Total %	10,0%	5,0%	10,0%	,0%	10,0%	35,0%
Total		Count	5	1	4	2	8	20
		Row %	25,0%	5,0%	20,0%	10,0%	40,0%	100,0%
		Column %	100,0%	100,0%	100,0%	100,0%	100,0%	100,0%
		Total %	25,0%	5,0%	20,0%	10,0%	40,0%	100,0%

Figure 2. Organization size x types of green energy practices adopted. Source: Authors' own research.

Organization size × Barriers

			Barriers			
			Financial	Technical expertise	Regulatory	Total
Organization size	Small	Count	3	2	2	7
		Row %	42,9%	28,6%	28,6%	100,0%
		Column %	30,0%	33,3%	50,0%	35,0%
		Total %	15,0%	10,0%	10,0%	35,0%
	Medium	Count	6	0	0	6
		Row %	100,0%	,0%	,0%	100,0%
		Column %	60,0%	,0%	,0%	30,0%
		Total %	30,0%	,0%	,0%	30,0%
	Large	Count	1	4	2	7
		Row %	14,3%	57,1%	28,6%	100,0%
		Column %	10,0%	66,7%	50,0%	35,0%
		Total %	5,0%	20,0%	10,0%	35,0%
Total		Count	10	6	4	20
		Row %	50,0%	30,0%	20,0%	100,0%
		Column %	100,0%	100,0%	100,0%	100,0%
		Total %	50,0%	30,0%	20,0%	100,0%

Figure 3. Barriers. Source: Authors' own research.

Chi-Square Tests

	Value	df	Asymptotic Sig. (2-tailed)
Pearson Chi-Square	10,10	4	,039
Likelihood Ratio	12,70	4	,013
Linear-by-Linear Association	,44	1	,505
N of Valid Cases	20		

Figure 4. Chi-Square tests. Source: Authors' own research.

One of the most encountered barrier when trying to adopt renewable resources in organizations is financial, especially in small and medium organizations underscoring the need for targeted benefits and grants, followed closely by lack of technical expertise (Figure 4).

The chi-square analysis revealed significant associations between organizational characteristics and barriers. Larger businesses reported less financial restrictions as a barrier to renewable energy implementation than smaller organizations. This implies that larger organizations frequently have greater access to money or internal resources for green initiative.

6. Future trends and innovations

Organizational green energy adoption is about to undergo a tremendous transition, fueled by rising technologies, fresh ideas, and global sustainability programs. The aim of these programs is to increase efficiency, availability, and incorporating renewable energy into organizational processes.

6.1 Emerging technologies: energy storage and smart grids

The unpredictability of solar and wind power is being addressed by new energy storage technologies, namely improved batteries (Wang et al., 2022). If more energy is generated than is consumed, these technologies serves to store this excess energy for later use during periods of lower generation. With the smart grid, using energy is even more intelligent since both supply and demand are always being adjusted to minimize waste and maximize energy use (Judge et al., 2022), which helps businesses tap into energy cost savings. These technologies are combined to facilitate the way for distributed and renewable energy system acceptance.

6.2 The role of AI and IoT in optimizing energy use

Energy consumption tracking and optimization in businesses is becoming increasingly important using machine learning (AI) and the Internet of Things (IoT). Energy demand can be forecasted, usage patterns can be modified, inefficiencies can be found, and the savings can be huge. Examples of IoT devices that can provide real time energy monitoring and control are intelligent thermostats, linked lighting systems (Ahmad and Zhang, 2021). AI and IoT join hands to build intelligent ecosystems where organizations are able to achieve sustainability goals while preserving operational efficiency.

6.3 Net-zero buildings and energy-positive offices

The next stage of sustainable construction is net-zero structures, the creation of as much power as we provide, and energy positive workplaces, creating more power than we consume (Mohammed et al., 2023). They are made with energy from renewable sources, energy efficient architecture and smart technologies, in order to minimize the environmental impact. These are reduced to solar panels, geothermal systems, and sophisticated insulating materials, amongst other things aimed at reducing energy demand while increasing productivity. Not only do they reduce carbon footprints, they are also powerful witnesses to the commitment that companies have put in place for the sustainability.

6.4 Global green initiatives and their impact

There are global organizational initiatives that are being influenced by the Paris Agreement (Ourbak and Magnan, 2018) the United Nations' Sustainable Development Goals (SDGs) (Gomez-Echeverri, 2018). These frameworks push organizations to match the effort of a global effort against climate change through the adoption of renewable energy, reduction of emissions and increase of energy efficiency. This has led to incentives by governments and business alliances, benchmarking, and collaboration (Florini and Pauli, 2018). These organisations which react proactively to these worldwide developments establish themselves as sustainability leaders, thus gaining competitive advantage and at the same time contribute to a greener environment.

By taking advantage of innovative technology and by cooperating with the rest of the world in its activities directed towards creating green initiatives, organizations could redevelop their energy strategy (Asif et al., 2020). These solutions do not only solve environmental problems but also offer avenues for prosperity, resilience of operations, and enhanced stakeholder participation.

7. Conclusion and recommendations

It is not a choice, but a necessity for companies that want to follow sustainability development objectives toreduce their carbon footprint and yield long term financial advantages. Switch to renewable energy sources like solar, wind and geothermal is very beneficial in many ways, including a lower carbon footprint, better operational efficiency, a better company image, and so on. Implementing green energy-related practices in organizations not only helps fight climate change, but can also help save money by improving usage of energy and the stability of future operations. This transformation also allows the companies to comply with regulatory criteria as well as with the wishes on a more eco-friendly stakeholder.

Organizations wishing to start their journey of renewable energy usage should take several critical measures:

1. *Conduct Energy Audits*: look for inefficiencies and find out about patterns. It helps the establishment of realistic, achievable sustainability targets.
2. *Invest in Renewable Technologies*: There are renewable energy solutions like solar panels, wind turbines, or even storage systems for energy—dependent on the geography and operating demands of the organization. These initiatives are not always possible without exploring ways to fund it, whether that be green financing or something from the government
3. *Partner with Experts*: Finally, collaboration between sustainability specialists, energy suppliers and technology specialists can help to simplify implementation and maximise the effect of green projects.
4. *Engage Employees*: Thus, training and integrating employees in sustainability activities can develop an environment friendly culture, not to mention a wonderful idea or better support for green energy projects.
5. *Integrate Solutions Gradually*: The approach of a phased transition between traditional and renewable energy systems minimises disturbance, which allows organizations to transition their infrastructure progressively.

Green energy will be playing an ever increasing role in organizational initiatives as we look towards the future. Technology will do the same for renewable energy, making it cheaper, more efficient. The campaign toward emission-free and energy-positive structures will gain traction as global accords and consumer expectations evolve. Organizations who take an active role in sustainability will profit not only economically, but also acquire a competitive advantage in an increasingly environmentally concerned market.

The foreseeable future of green energy is likely to witness further acceptance, thanks to the collaborative efforts of governments, business, and individual consumers. Organizations that proactively adopt these adjustments are going to contribute to a sustainable environment and establish themselves as pioneers in the emerging age of energy management.

References

Adeniyi, A. G., Amusa, V. T., Iwuozor, K. O. and Emenike, E. C. (2022). Thermal recycling strategy of Coca-Cola PVC label films by its co-carbonization with Terminalia ivorensis leaves. Cleaner Engineering and Technology 11: 100564. https://doi.org/10.1016/j.clet.2022.100564.

Aeknarajindawat, N. S. B. S. P. (2020). The impact of natural resources, renewable energy, economic growth on carbon dioxide emission in Malaysia. International Journal of Energy Economics and Policy 10.

Ahmad, T. and Zhang, D. (2021). Using the internet of things in smart energy systems and networks. Sustainable Cities and Society 68: 102783. https://doi.org/10.1016/j.scs.2021.102783.

Aien, M. and Mahdavi, O. (2020). On the way of policy making to reduce the reliance of fossil fuels: case study of Iran. Sustainability 12(24): 10606. https://doi.org/10.3390/su122410606.

Al-Shetwi, A. Q. (2022). Sustainable development of renewable energy integrated power sector: Trends, environmental impacts, and recent challenges. Science of The Total Environment 822: 153645. https://doi.org/10.1016/j.scitotenv.2022.153645.

Asif, M. S., Lau, H., Nakandala, D., Fan, Y. and Hurriyet, H. (2020). Adoption of green supply chain management practices through collaboration approach in developing countries – From literature review to conceptual framework. Journal of Cleaner Production 276: 124191. https://doi.org/10.1016/j.jclepro.2020.124191.

Beardsmore, G. (n.d.). Hot Dry Rocks Pty Ltd DIRECT USE OF GEOTHERMAL ENERGY: GLOBAL MARKET SCAN. www.hotdryrocks.com.

Choudhary, P. and Srivastava, R. K. (2019). Sustainability perspectives—a review for solar photovoltaic trends and growth opportunities. Journal of Cleaner Production 227: 589–612. https://doi.org/10.1016/j.jclepro.2019.04.107.

Costea, G., Pusch, M. T., Bănăduc, D., Cosmoiu, D. and Curtean-Bănăduc, A. (2021). A review of hydropower plants in Romania: Distribution, current knowledge, and their effects on fish in headwater streams. Renewable and Sustainable Energy Reviews 145: 111003. https://doi.org/10.1016/j.rser.2021.111003.

Elkhatat, A. and Al-Muhtaseb, S. (2024). Climate change and energy security: a comparative analysis of the role of energy policies in advancing environmental sustainability. Energies 17(13): 3179. https://doi.org/10.3390/en17133179.

Florini, A. and Pauli, M. (2018). Collaborative governance for the Sustainable Development Goals. Asia & the Pacific Policy Studies 5(3): 583–598. https://doi.org/10.1002/app5.252.

Gökgöz, M. T. G. F. (2018). Energy security and renewable energy efficiency in EU. Renewable and Sustainable Energy Reviews 96: 226–239. https://doi.org/https://doi.org/10.1016/j.rser.2018.07.046.

Gomez-Echeverri, L. (2018). Climate and development: Enhancing impact through stronger linkages in the implementation of the Paris Agreement and the Sustainable Development Goals (SDGs). In Philosophical Transactions of the Royal Society A: Mathematical, Physical and Engineering Sciences (Vol. 376, Issue 2119). Royal Society Publishing. https://doi.org/10.1098/rsta.2016.0444.

Guelpa, E., Bischi, A., Verda, V., Chertkov, M. and Lund, H. (2019). Towards future infrastructures for sustainable multi-energy systems: A review. Energy 184: 2–21. https://doi.org/10.1016/j.energy.2019.05.057.

Hassan, Q., Viktor, P. J., Al-Musawi, T., Mahmood Ali, B., Algburi, S., Alzoubi, H. M. et al. (2024). The renewable energy role in the global energy transformations. Renewable Energy Focus 48: 100545. https://doi.org/10.1016/j.ref.2024.100545.

Hu, Y., Bai, W., Farrukh, M. and Koo, C. K. (2023). How does environmental policy uncertainty influence corporate green investments? Technological Forecasting and Social Change 189: 122330. https://doi.org/10.1016/j.techfore.2023.122330.

Ibrahim Adedeji Adeniran, Angela Omozele Abhulimen, Anwuli Nkemchor Obiki-Osafiele, Olajide Soji Osundare, Edith Ebele Agu and Christianah Pelumi Efunniyi. (2024). Strategic risk management in financial institutions: Ensuring robust regulatory compliance. Finance & Accounting Research Journal 6(8): 1582–1596. https://doi.org/10.51594/farj.v6i8.1508.

Janiszewska, D. and Ossowska, L. (2022). The role of agricultural biomass as a renewable energy source in European Union Countries. Energies 15(18): 6756. https://doi.org/10.3390/en15186756.

Judge, M. A., Khan, A., Manzoor, A. and Khattak, H. A. (2022). Overview of smart grid implementation: Frameworks, impact, performance and challenges. Journal of Energy Storage 49: 104056. https://doi.org/10.1016/j.est.2022.104056.

Kabeyi, M. J. B. and Olanrewaju, O. A. (2022). Sustainable energy transition for renewable and low carbon grid electricity generation and supply. Frontiers in Energy Research 9. https://doi.org/10.3389/fenrg.2021.743114.

Khan, F. A., Pal, N. and Saeed, Syed. H. (2018). Review of solar photovoltaic and wind hybrid energy systems for sizing strategies optimization techniques and cost analysis methodologies. Renewable and Sustainable Energy Reviews 92: 937–947. https://doi.org/10.1016/j.rser.2018.04.107.

Lacal-Arántegui, R. (2019). Globalization in the wind energy industry: contribution and economic impact of European companies. Renewable Energy 134: 612–628. https://doi.org/10.1016/j.renene.2018.10.087.

Lu, Y., Khan, Z. A., Alvarez-Alvarado, M. S., Zhang, Y., Huang, Z. and Imran, M. (2020). A critical review of sustainable energy policies for the promotion of renewable energy sources. Sustainability 12(12): 5078. https://doi.org/10.3390/su12125078.

Mahalakshmi, S., Nallasivam, A., Kumar, S. H., Kalavathy, K. S. and Desai, K. (2024). Sustainability by Design: IKEA's Eco-Friendly Marketing Efforts (pp. 145–161). https://doi.org/10.2991/978-94-6463-374-0_14.

Mohammed, G. A., Mabrouk, M., He, G. and Abdrabo, K. I. (2023). Towards sustainable cities: a review of zero energy buildings techniques and global activities in residential buildings. Energies 16(9): 3775. https://doi.org/10.3390/en16093775.

Moorthy, K., Patwa, N., Gupta, Y., Rahman, A., Campus, K. and P Jain, M. S. (2019). Breaking barriers in deployment of renewable energy. Heliyon e01166. https://doi.org/10.1016/j.heliyon.2019.

Ourbak, T. and Magnan, A. K. (2018). The Paris Agreement and climate change negotiations: Small Islands, big players. pp. 2201–2207. In Regional Environmental Change (Vol. 18, Issue 8). Springer Verlag. https://doi.org/10.1007/s10113-017-1247-9.

Pakulska, T. and Poniatowska-Jaksch, M. (2022). Digitalization in the renewable energy sector—new market players. Energies 15(13): 4714. https://doi.org/10.3390/en15134714.

Paulillo, Andrea, Aleksandra Kim, Christopher Mutel, Alberto Striolo, Christian Bauer and Paola Lettieri. (2021). Influential parameters for estimating the environmental impacts of geothermal power: A global sensitivity analysis study. Cleaner Environmental Systems 3.

Peduzzi, P. (2019). The disaster risk, global change, and sustainability Nexus. Sustainability 11(4): 957. https://doi.org/10.3390/su11040957.

Poudel, B., Maley, J., Parton, K. and Morrison, M. (2021). Factors influencing the sustainability of micro-hydro schemes in Nepal. Renewable and Sustainable Energy Reviews 151: 111544. https://doi.org/10.1016/j.rser.2021.111544.

Prăvălie, R., Sîrodoev, I., Ruiz-Arias, J. and Dumitraşcu, M. (2022). Using renewable (solar) energy as a sustainable management pathway of lands highly sensitive to degradation in Romania. A countrywide analysis based on exploring the geographical and technical solar potentials. Renewable Energy 193: 976–990. https://doi.org/https://doi.org/10.1016/j.renene.2022.05.059.

Prehoda, E., Pearce, J. and Schelly, C. (2019). Policies to overcome barriers for renewable energy distributed generation: a case study of utility structure and regulatory regimes in Michigan. Energies 12(4): 674. https://doi.org/10.3390/en12040674.

Rahman, A., Farrok, O. and Haque, M. M. (2022). Environmental impact of renewable energy source based electrical power plants: Solar, wind, hydroelectric, biomass, geothermal, tidal, ocean, and

osmotic. Renewable and Sustainable Energy Reviews 161: 112279. https://doi.org/10.1016/j.rser.2022.112279.
Song, W., Ren, S. and Yu, J. (2019). Bridging the gap between corporate social responsibility and new green product success: The role of green organizational identity. Business Strategy and the Environment 28(1): 88–97. https://doi.org/10.1002/bse.2205.
Stanitsas, M. and Kirytopoulos, K. (2023). Sustainable energy strategies for power purchase agreements (PPAs). Sustainability 15(8): 6638. https://doi.org/10.3390/su15086638.
United Nations Climate Change. (n.d.). Retrieved March 20, 2023, from https://unfccc.int/.
Vanegas Cantarero, M. M. (2020). Of renewable energy, energy democracy, and sustainable development: A roadmap to accelerate the energy transition in developing countries. Energy Research & Social Science 70: 101716. https://doi.org/10.1016/j.erss.2020.101716.
Villca-Pozo, M. and Gonzales-Bustos, J. P. (2019). Tax incentives to modernize the energy efficiency of the housing in Spain. Energy Policy 128: 530–538. https://doi.org/10.1016/j.enpol.2019.01.031.
Wang, W., Yuan, B., Sun, Q. and Wennersten, R. (2022). Application of energy storage in integrated energy systems—A solution to fluctuation and uncertainty of renewable energy. Journal of Energy Storage 52: 104812. https://doi.org/10.1016/j.est.2022.104812.
Worku, M. Y. (2022). Recent advances in energy storage systems for renewable source grid integration: a comprehensive review. Sustainability 14(10): 5985. https://doi.org/10.3390/su14105985.
Ziolo, M., Filipiak, B. Z., Bąk, I. and Cheba, K. (2019). How to design more sustainable financial systems: the roles of environmental, social, and governance factors in the decision-making process. Sustainability 11(20): 5604. https://doi.org/10.3390/su11205604.
ZOU, C., Xiong, B., Xue, H., Zheng, D., Ge, Z., Wang, Y. et al. (2021). The role of new energy in carbon neutral. Petroleum Exploration and Development 48(2): 480–491. https://doi.org/10.1016/S1876-3804(21)60039-3.

Chapter 9

Closing the Loop
Practical Innovations in Fashion Waste Recycling

Eugen Constantin Ailenei, Alexandra Bodoga, Andreea Nistorac, Dorin-Savin Ionesi* and *Maria Carmen Loghin*

1. Introduction

Sustainability is the characteristic of an activity to be carried out without exhausting natural resources and without degrading the environment. Establishing a balance between environmental protection, economic growth and finding alternative resources are decisive factors for ensuring the sustainable development of human society. The principles of the circular economy consist of the reduction of resource waste and the use of natural resources through responsible consumption as well as the recovery of waste materials, extending the life of products and promoting sustainable design that facilitates the repair and modernization of products, recycling and regeneration by reintroducing materials to the economic cycle through and the promotion of sustainable business models (Avadanei, 2021; Nistorac, 2024).

In the last few decades, population growth worldwide has accelerated, due to migration movements, social transformations and political and economic circumstances. This trend will continue in the coming decades, with the world population expected to grow from 8.1 billion today to approximately 9.8 billion by 2050 with 70% of the population concentrated in large metropolitan centers.

" Gheorghe Asachi" Technical University of Iasi-Romania, Faculty of Industrial Design and Business Management, No. 29 Prof. D. Mangeron Blvd., 700050, Iasi, Romania.
Email: maria-carmen.loghin@academic.tuiasi.ro
* Corresponding author: constantin-eugen.ailenei@academic.tuiasi.ro

DOI: 10.1201/9781003518990-9

As a result, resources are becoming scarcer and climate change is a certainty with devastating effects in all fields (ONU, 2024).

Worldwide, the textile industry is considered one of the most polluting industries (Nørup et al., 2018). The European Union considers clothing products and household textile products highly polluting, and it is necessary to adopt specific policies to reduce the impact on the environment (Beton et al., 2019). With the increase in the standard of living and its achievement by countries, such as China, there is a sharp increase in the consumption of textile products. This trend is not necessarily due to demographic growth alone, but rather a mix of factors that includes the level of prosperity of the population as well as the Western (European and American) consumption model (Nørup et al., 2018).

Fashion industry is the main and largest consumer of textile fibers (Textile Exchange, 2022), it promotes a model of rapid consumption of various clothing items, all of which are deliberately designed with a short built-in lifespan. Concretely, in the clothing sector, the so-called fast fashion trend (fast fashion) develops through the regular replacement of articles, whose availability and accessibility determine the increase in the quantities of textile materials used (Boone, 2009). Along with other fast-growing major waste streams such as e-waste, the volume of waste from the fashion industry is expected to grow from 92 million tonnes in 2015 to 148 million tonnes in 2030 (Truscott et al., 2017).

The manufacture of textile materials involves the use of virgin raw materials, these resources being at a critical stage of availability due to the depletion of natural resources. Considering the entire life cycle of textile products, from production to disposal in landfills of pre- and post-consumer items at the end of the life cycle, "textile production contributing 10% of carbon emissions, is the second the most polluting sector in the world and represents a complex and challenging waste stream" (Bell et al., 2018).

Statistical data reveal that 70% of the world's textile fiber consumption corresponds to synthetic fibers, made from petroleum-derived polymers such as nylon, acrylic, polyester and polypropylene. From fiber cultivation and synthesis, to yarn and fabric manufacturing, the textile industry uses significant natural and energy resources with a substantial environmental footprint (Muthu et al., 2012b; Rochman et al., 2015).

Textile waste streams include pre-consumer waste (fibre and yarn waste, cutting waste and degraded textile articles), post-consumer (end-of-life textile articles) and industrial waste. For example, the European Union (EU) generates approximately 5.8 million tons of post-consumer textile waste annually, of which only 25% of the total volume is recycled into low-value products or through incineration, and the remaining 75% is directed to landfills of garbage (Briga-Sá et al., 2013).

2. Some limitation in processing of textile waste

The main environmental impact factors of: virgin fiber production, global consumption of textile products and the high percentage of disposal in landfills or incineration confirm the advantage of recycling and extending the life of textiles (Muthu et al., 2012a). On the other hand, the technologies that used to transform textile waste into energy, through incineration, pyrolysis and gasification, had an inalienable role in significantly reducing the mass and volume of waste, as well as the recovery of energy from non-recyclable materials and their complete disinfection (Arena, 2012).

According to the MFF Report (Elander and Ljungkvist, 2016) the main problems regarding textile waste recycling are:

- Insufficient collection and sorting due to the lack of economically viable recycling technologies and the limitations of mechanical, chemical recycling and sorting technologies (Mcgregor, 2015; Ouyang and Cao, 2012).
- Difficulties in recycling fabrics with mixed fiber composition (cotton/polyester, nylon/elastane, etc.), due to the varied material composition, heterogeneity and interconnection of fibers (Marconi et al., 2018; Hawley, 2006).
- Difficulties with the removal of dyes, additives and processing agents from textile materials, some of which are toxic and carcinogenic (Bhuiyan et al., 2015).
- Economic aspects regarding the low viability of sorting, recovery of recycled fibers and recycling itself. The economic feasibility compared to the cost of producing virgin fibers is influenced by the costs of recycling, collecting and transporting textile waste (SEPA, 2015).
- The unlikely possibility of changing the upward trend in textile waste generation without a change in consumer behavior (Nixon and Saphores, 2009; Conke, 2018). Delaying textile waste collection policies and standards has a negative impact on recycling efficiency (Payne, 2015).

The reuse of textile products refers to different ways of extending the useful life, with or without prior modification, and transferring it to a new owner. This method requires low energy consumption, as it does not require special resources and is usually limited to collection and resale. In the case of clothing, the energy used to collect, order and resell old clothes can be 10–20 times less energy than recycling. The main ways in which textiles are reused are: through thrift stores, flea markets, donations and sales to charities (Ben-Eli, 2018). Considering the number of items that consumers throw away every day, the reusing process holds great potential.

For the transfer of post-consumer textile waste to new users, whether it is to extend their life cycle or to transform them into new products, they must comply with a certain biological safety standard. Traditional cleaning methods, such as water-based washing, may not be suitable for all types of textile waste and may require significant energy and water consumption. Textiles are essential materials in many industries, including healthcare and food service, where cleanliness is essential to prevent the spread of infection. However, they can easily be contaminated with bacteria, viruses, or other harmful microorganisms that can cause disease. Several methods of textile disinfection can be discussed, including chemical, thermal, and physical techniques and their effectiveness in eliminating microorganisms. However, most of these disinfection methods result in harmful chemical residues that need to be neutralized or the costs associated with using them are too high and thus become economically inefficient.

Thus, through European patent application number EP24177858, a "Device and method for disinfection of contaminated clothing products" is proposed. This invention, which provides a simple device for the disinfection of clothing products contaminated with pathogens, consist of a gas-impermeable enclosure made of plastic material equipped with a sealing system wherein the gas-impermeable enclosure has a one-way mechanical valve connectable to a vacuum pump. The said mechanical valve is provided with a closing cap wherein the gas impermeable enclosure further comprises an input connectable to an ozone generator. The invention also includes a method of disinfecting articles of clothing contaminated with pathogens using the described device. The process involves placing the contaminated products in a gas-tight enclosure, removing air to create a vacuum, and then introducing ozone at a controlled rate for a predetermined duration.

The advantages of the device and method provided in this invention are:

- The disinfection process is a dry process being made with ozone gas;
- the disinfectant, namely the ozone (O_3) gas, is consumed without leaving any residues that need to be neutralized later on;
- by vacuuming and removing the air from the gas-impermeable enclosure, between the textile products, as well as inside the structures from which the textile products are made, it allows ozone to act effectively on pathogens regardless of their positioning on the product to be disinfected;
- the quantity of ozone is minimal because the removal of the air from the gas-impermeable enclosure results in the direct action of the ozone on the clothing/textile to be disinfected; also, the ozone has a specific flow rate and a long time of action until it is consumed;
- there is the possibility to disinfect different quantities of clothing according to the necessity and the volume of the gas-impermeable enclosure;
- during disinfection, their odors are also carried out.
- the invention also allows for the repeated use of the device and method by sealing each gas-impermeable enclosure after vacuuming and ozone injection, and then repeating the process with a new enclosure containing a fresh batch of clothing items to be disinfected;
- the disinfection method and device are simple, and easy to use, with very high disinfection rates and low costs;
- also, it does not use any chemical agents that need to be deactivated/neutralized by the end of the process;
- it has reproducible results with very high disinfection rates.

The process and method of disinfection of clothing products contaminated with pathogens using ozone allows the safe reuse of post-consumer waste both as second-hand products and as raw materials for the creation of new products (Bodoga et al., 2023).

Currently the textile waste recycling industry is still in its early development stages, its recovery as a raw material for new applications is limited by a distinct gap between current recycling technology and the profitability of commercial-scale manufacturing of new products. The ideal route is the one in which the textile waste is first reused, after which it will enter mechanical and/or chemical recycling processes until reaching the lower threshold of the economic reliability of the fibers, polymers and/or oligomers thus obtained. The new materials obtained can be used to make similar products. The limitations are mainly due to the complexity of the blends, the limited recycling cycles (synthetic fibers 7–9 cycles, cellulose fibers 4–6 cycles) after which their length becomes too short to be suitable for use, as well as the high load of over 800 chemical substances used in textile processing (dyes, softeners, fire retardants, preservatives, etc.) (Oecotextiles, 2016; KEMI, 2014).

The closed-loop waste recycling approach with the aim of obtaining fibers to replace the virgin raw material in a 1:1 ratio may prove to be an unprofitable solution due to the qualities of the recovered polymer or fibers, the consumption of substances and energy required for the recycling processes (Sandin and Peters, 2018; Ekström and Salomonson, 2014). Closely related to the quality of the recycled raw material and the products that can be obtained, recycling is classified into upcycling or downcycling (plus value or minus value) with a major impact on the decision of the optimal recycling route.

In general, the recycling of textile waste is of the downcycling type. The transformation of unsorted textile waste that is not preferred for other recycling scenarios into innovative products for the construction sector can be considered upcycling, as this waste will acquire a new value in a market segment where demand exceeds supply.

At the same time, the construction industry is another sector with a significant impact on the environment, consuming natural resources globally at an accelerated rate and contributing to greenhouse gas emissions, especially through the use of traditional building materials, such as cement, brick, autoclaved cellular concrete (BCA), polystyrene or mineral wool.

There is, therefore, an urgent need to find sustainable solutions to address these two major issues: managing textile waste and reducing the negative impact of construction materials on the environment.

Extending the life cycle of textile materials by recycling and reusing them for the development of innovative building materials contributes not only to the efficient management of textile waste, but also to the development of sustainable materials that focus on the use of textile waste in combination with materials such as polypropylene or expanded polystyrene, thus providing a viable alternative to traditional materials. The circular economy is a global priority, and extending the life cycle of textile products aligns with this trend. The transformation of textile waste into construction materials supports the transition to a circular economic model, where resources are reused, repaired and recycled, thus minimizing the consumption of new resources and the impact on the environment. By extending the life cycle of these materials, the research proposes innovative solutions to reduce the amount of textile waste that ends up in landfills or is incinerated.

The development of such solutions represents an important step towards achieving a circular economy, where resources are used to their maximum capacity and the impact on the environment is reduced. In this context, the nature of innovation is multidisciplinary, combining elements from the field of materials science, construction engineering and environmental protection.

3. Opportunity for new innovative products

In order to create materials that use textile waste with complex fibrous composition, intended to be used in construction, the identification of the main types of textile waste plays an important role. To have a better picture of the amount and composition of waste generated by the textile industry, the demand was analyzed both by fiber type (natural and chemical) and by the proportion of each type of fiber in this industry. In 2018, the amount of polyester fibers consumed worldwide far exceeded the other types of chemical fibers, but also those of natural fibers. Figure 1 shows that the share of synthetic fibers is approximately 62.3%, and cotton represents 24.4% of demand. Of the total consumption of synthetic fibers, polyester has more than 80% of the total demand. The percentage of fibers used for clothing has a significant share of over 55% (Opperskalski et al., 2019).

The statistical analysis carried out helped us to identify the main category of textile waste that is found in abundance in pre and post-consumer waste from the point of view of fibrous composition. These are generally cotton-type fabrics characterized by complex fibrous compositions, but incorporating polyester in different proportions mixed with natural and/or synthetic fibers, with different mass densities and colors, unsorted.

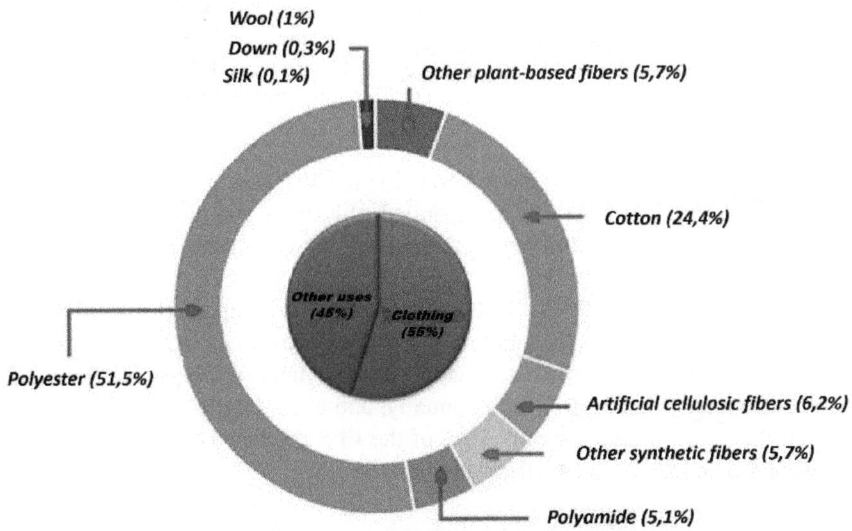

Figure 1. The market's shares of textile fibers.

The identification of the market niche of the new products and the establishment of the necessary physical-mechanical characteristics were decided after studying the categories of products that already exist in the construction materials market, the size of the sales markets, the degree of occupation of the production capacities and the price of the raw materials necessary to satisfy the request.

The two solutions developed are complementary for the recovery of textile waste in the construction industry. The proposed technology for making composite materials using the tandem of textile waste and polypropylene matrix is more suitable for structural applications due to the superior mechanical properties of the composite material, while the material variant made from textile waste and expanded polystyrene focuses on energy efficiency and sustainability by developing a lightweight and ecological insulating material.

An important element in making composite materials is the choice of the matrix. In the case of composite material for structural applications such as composite building panels, a 100% polypropylene thermoplastic matrix was selected. For economic and ecological reasons, two types of polypropylene waste were taken into account, namely bio-oriented polypropylene waste and polypropylene non-woven material waste. The decision to use polypropylene as a matrix was based on the following considerations:

- ***Abundance*** of polypropylene waste. Bi-oriented polypropylene film is frequently used in the packaging industry with the most diverse applications (food packaging, textile packaging, etc.), polypropylene non-woven materials are used to make disposable products (overalls, protective masks, disposable sheets, etc.). Medical consumables used in the context of the COVID-19 pandemic generate significant amounts of pre- and post-consumer waste.
- ***High degree of recyclability***. Polypropylene polymer is particularly thermally stable and therefore suitable for multiple recycling. Even after 15 consecutive recyclings processes, PP did not significantly change its tensile and impact resistance compared to the original material. The material viscosity obtained after the 15th extrusion was almost identical to that after the first extrusion sequence (Torney, 2018).

The development of the new material for the thermal insulation of buildings is based on textile waste and uses expanded polystyrene as a matrix. This is an extremely durable material, but one that breaks down very hard in the environment. EPS can persist in the environment for hundreds of years. For example, studies suggest that EPS can take 500 to 1,000 years to fully decompose (EPA, 2021). EPS breaks down into smaller pieces, but does not break down completely, contributing to the accumulation of microplastics in the environment. Globally, approximately 14 million tons of EPS are produced annually, and according to a report by Plastics Europe (2021), approximately only 10% of the EPS produced annually is recycled, with the rest being sent to landfills or incinerated.

Both solutions use textile waste to develop innovative construction solutions, but the approaches differ significantly:

- RECOTEX FRC composite intended as an alternative to wood-based boards used in construction and the furniture industry emphasizes mechanical stiffness by using a strong thermoplastic matrix combined with unsorted textile reinforcement.
- TermoTEX prioritizes energy efficiency by developing an insulating material based on recycled cotton and expanded polystyrene.

These differences reflect both the objectives pursued (mechanical vs. thermal) and the adaptability of each method to the specific requirements of the construction industry.

3.1 FRC RECOTEX

Based on the opportunity studies carried out, it was found that the global demand for wood products is continuously expanding, being determined by the combination of population growth and economic growth, and consumption is influenced by technological developments and environmental considerations. Countries respond to these global changes in several ways: primarily by increasing imports or changing the resource base, through an expansion of planting resources. Plantations have grown from 18 million hectares in 1980 to over 200 million hectares today, and cultivated timber now accounts for over 40% of the industrial timber supply. According to the source, industry-wide capacity utilization was 88% in 2016, and the forecast is for growth in 2017. In fact, in the second quarter presentation, some of the manufacturers (Norbord) reported a capacity utilization rate production of 99% (Koenhemsi, 2017).

Due to these reasons, the development of products that can be adopted as alternatives to thin, long and oriented OSB-type chipboards represents a promising initiative, which addresses both the environmental problems of the textile industry and the economic and performance needs of the construction sector.

3.1.1 Characterization of composite material reinforcements

The fibrous composition of the textile waste used as reinforcement was identified according to the existing product sheets, and where this was not possible, the fibrous composition was identified through specific test methods. Sorting into categories of fiber types is a difficult operation, the main reason why unsorted waste is difficult to recycle. The following categories of cotton-type fabrics with different fibrous mixtures and the approximate percentages of participation in the total mass of textile waste were identified, which were used to make FRC RECOTEX boards:

- 40%–100% polyester fabrics,
- 20%–97% polyester and 3% elastane fabrics,
- 15%–88% polyester and 12% elastane fabrics,
- 10% - fabrics from 82% polyester, 16% viscose and 3% elastane,
- 15% - fabrics with more than 70% polyester and other fibers.

The reinforcement used comes from post-consumer textile waste from the fashion industry's cutting department. Over 35% of pre-consumer waste is generated in the primary processing phase of the raw material, of which 9% in fiber production and 91% in the technological operations of product manufacturing (Global Fashion Agenda and The Boston Consulting Group, 2017).

3.1.2 Characterization of the composite material matrix

To make the composite material, the matrix from technological waste from the range of thermoplastic polymers was chosen. These are presented in Table 1.

Table 1. Thermoplastic matrix used.

Matrix		**Features**	**Provenance**
	Bi-oriented polypropylene film. (BOPP)	Thickness 30, 35 and 40 microns. Melting point 165°C	Bi-oriented polypropylene film waste from the packaging industry
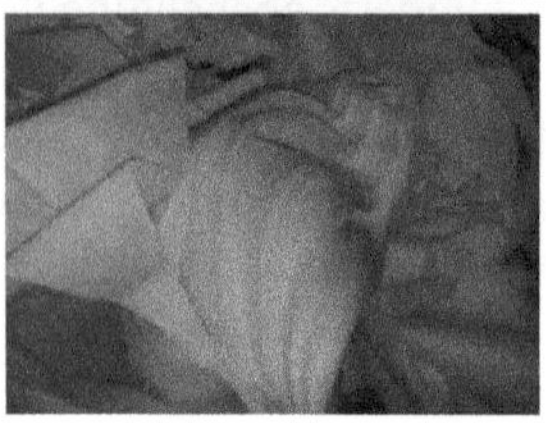	100% polypropylene non-woven material obtained by calendering.	Mass density between 75 g/m^2 and 100 g/m^2. Melting point 165°C	Waste resulting from the cutting operation in the upholstery section of the furniture industry

3.1.3 Technological flow

The matrix/reinforcement mixture was fed to the cutting machine in two passes in order to achieve cutting on both axes of the fed textile waste. Thus, over 66% of the strips obtained have dimensions in the range of 5–12 cm^2. This cutting is suitable from the perspective of use within an industrial technological flow for the preliminary processing of textile waste and its mixing with the matrix, necessary to make composite boards, due to the large quantities processed and the use of a unique reinforcing/matrix mix recipe. Figure 2 shows the version of the technological flow.

Thermoforming was performed on a machine specially designed and manufactured for the experiments from the Research Center of the Faculty of Industrial Design and Business Management, TUIASI. The thermoforming machine, shown in Figure 3, has two plates that can be electrically heated in the range 0–300°C to melt the polypropylene matrix, and cooled in an open circuit, with water, to reduce the solidification time of the polypropylene.

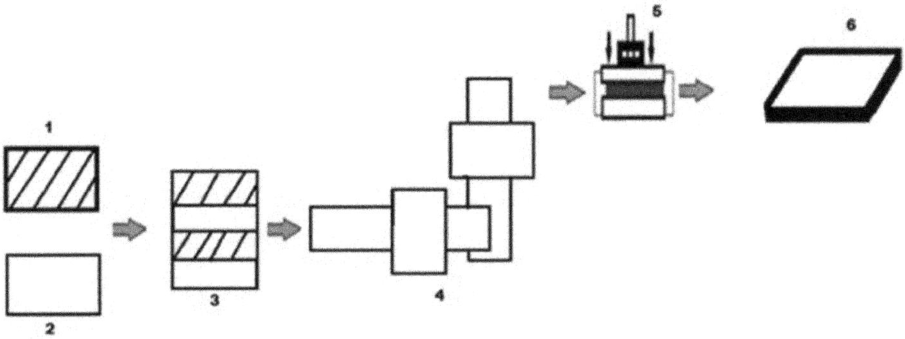

Figure 2. Technological flow for producing FRC RECOTEX boards. 1 - matrix; 2 - reinforcing; 3 - matrix/reinforcer mixing; 4 - aggregate cutting machines; 5 - thermo pressing machine; 6 - composite material.

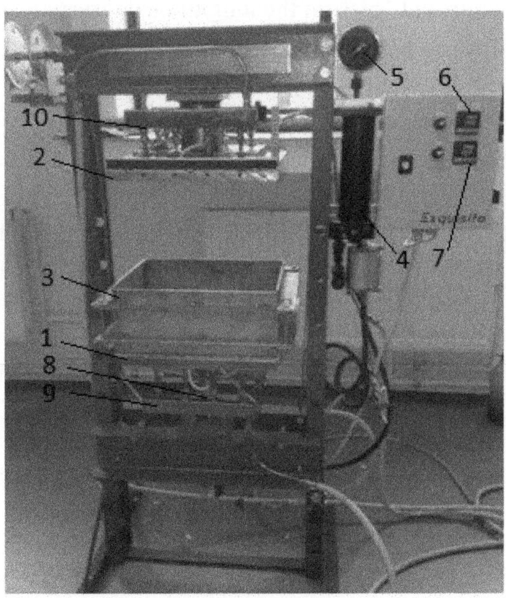

Figure 3. Thermoforming machine. Legend: 1 - lower plate; 2 - upper plate; 3 - stainless steel mold; 4 - hydraulic pump; 5 - manometer; 6,7 - upper/lower platen electronic temperature regulator; 8 - hydraulic cylinder; 9, 10 - open circuit water cooling system.

The thermoforming device is equipped with a mold located on the lower platen which is set in motion by a pneumatically actuated hydraulic cylinder that can act with a pressing force ranging between 0÷30 US-To.

3.1.4 Mechanical properties of the composite material

The mechanical properties of the obtained waste-based composites were evaluated in terms of stiffness and tensile strength using the LBG TC10 testing machine equipped with a 10 kN load cell and equipped with TCSoft 2004 Plus software. Three-point bending strength and modulus of elasticity tests were performed according to BS EN 310: 1996 (SR EN, 1996). This standard for wood-based materials was used because the investigated composite materials are intended to replace oriented strand board in construction applications. The SR EN 300: 2006 standard (SR EN, 2006) was used to compare the mechanical properties of OSB/3 boards in the 6–10 mm thickness range with composite boards obtained.

Through the practical validation of the oprim variant and the comparison of the results obtained after determining the physical-mechanical characteristics of the composite material obtained with those of the OSB plates, we can conclude that this solution is a viable one. Thus, the obtained results validate the following aspects in comparison with OSB/3 boards:

- The tensile strength of the waste-based composites is higher by 1030.97% in the transverse direction and 410.21% in the longitudinal direction.
- The flexural strength of waste-based composites is higher by 351.92% in the transverse direction and 150% in the longitudinal direction.
- The stiffness of waste-based composites is 8% lower in the transverse direction and 60% lower in the longitudinal direction.
- The density of the composite is 173% higher.
- The values obtained for thickness swelling (Gt) and water absorption (A) indicate that the composite material is hydrophobic.
- The composite material has a coefficient of thermal conductivity (λ) of 0.24 W/mk (λ hardwood = 0.23 W/mk)
- The recycling scenario was validated by making a plate from 100% cut and thermo-pressed composite material waste. The values obtained as a result of the determinations carried out according to the standards confirmed the possibility of recycling and obtaining composite material boards with characteristics comparable to those of OSB boards.

The Figure 4 shows the comparative values of the mechanical characteristics obtained for the plates related to the optimal variant of composite material, the OSB plates in transverse and longitudinal section, but also the plates made of recycled composite material are presented.

The obtained results reveal that the development of composite material panels for construction applications made from pre-consumer textile material waste, in order to replace the traditional wood-based composite boards of the OSB type, is a viable solution.

European patent no. EP3871853 entitled "Thermopressed composite material with polypropylene waste matrix reinforced with textile waste with multiple fibrous composition, as well as method and device for its production" was obtained with funding from the Ministry of Research and Innovation within the TUIASI-COMPETE Project of the "Gheorghe Asachi" Technical University of Iaşi. A series of prototype products, as shown in Figure 5, were also developed and produced.

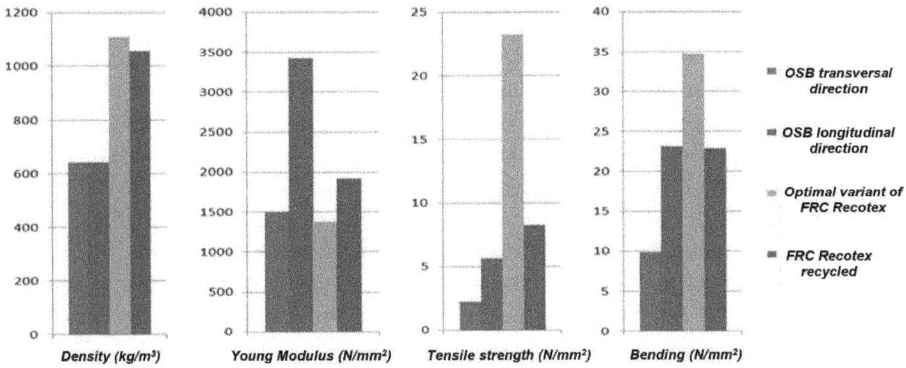

Figure 4. Comparative graphic representation of the mechanical characteristics related to the investigated plates.

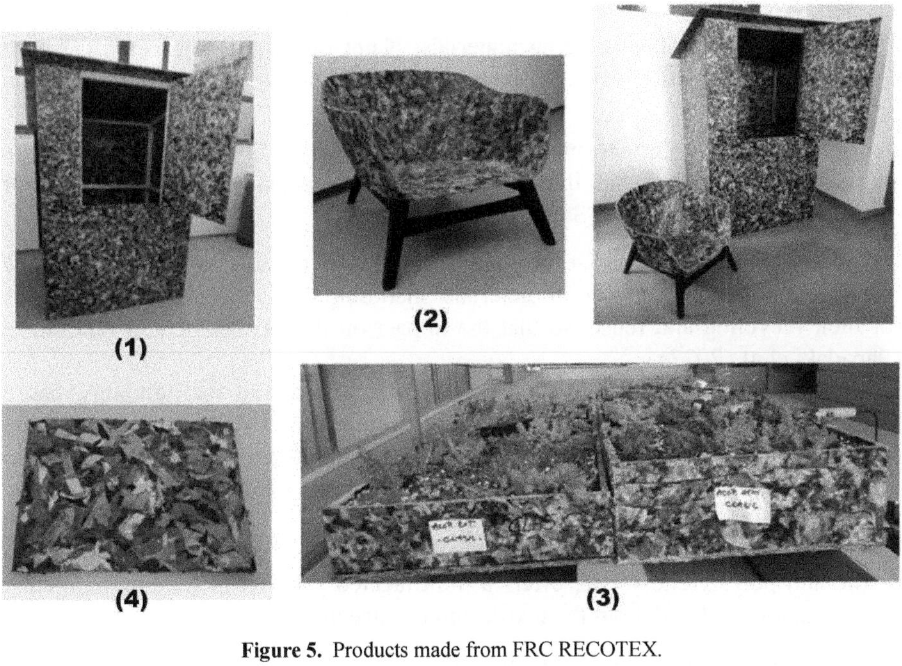

Figure 5. Products made from FRC RECOTEX.

Composite panels were produced under industrial conditions and used to build a demountable garden shed (1). The 3D thermoforming hypothesis was also tested by making an armchair (2). Other product ideas were also explored, such as green roofs (3) and decorative board (4). Due to the versatility of the composite material and its physical and mechanical properties, the applications it can incorporate are virtually unlimited.

3.2 TermoTEX

Modern building codes specify the needs of structural engineering systems and limit the thermal performance of the building envelope, which also determines the best possible use of energy under certain technical and financial circumstances.

Technical solutions based on increasing energy efficiency promote sustainability, reducing operational costs and environmental impact, while improving the flexibility, durability and resilience of buildings. One focus is thermal protection to improve the building envelope through thermal insulation (Makrygiannis and Karalis, 2023).

Insulating materials such as mineral wool, polystyrene and other insulating foams are essential for the energy performance of buildings, but the generation of waste from their use raises significant environmental problems. These materials often take a long time to decompose and can create pollution problems, especially if they are not managed properly. Major challenges include:

- *Difficulty of recycling*: Some insulating materials are difficult to recycle or do not have technologies available to recycle them efficiently.
- *Improper disposal*: Insulating materials, if not disposed of properly, can lead to the accumulation of waste in landfills, contributing to long-term pollution problems.
- *Health impact*: Certain insulation materials, such as fiber-based ones, can be harmful to human health if handled improperly, especially during demolition or renovation of buildings (Superti et al., 2021)

These aspects emphasize the need for more circular approaches in the management of waste insulation materials, including the development of solutions for their recycling and reuse, so that the impact on the environment is minimized (Papadaki et al., 2022).

The development of a new thermal insulation material for buildings, based on textile waste and expanded polystyrene, represents a significant opportunity to improve energy efficiency and protect the environment. This type of material combines two sources of waste that, through recycling and processing, can become valuable resources for the construction industry.

This approach, which involves developing a novel insulating material from textile and polystyrene waste, offers a multifaceted solution to the environmental impacts associated with both the textile and construction industries:

- Reducing the amount of textile waste;
- Use in a new life cycle as long as possible;
- Reducing the amount of expanded polystyrene waste;
- Development of innovative insulation materials that can help meet NZEB requirements.

3.2.1 Characterization of thermal insulation material reinforcements

The textile waste was picked up chopped and mixed, and its composition was identified according to the related production during the collection period. The composition of the mixture of post-consumer textile waste used is over 80% cotton mixed with various percentages of synthetic and artificial fibers.

The mixture of shredded textile waste is inhomogeneous, both in terms of composition, although it consists mostly of cotton, and especially in terms of shredding size (Figure 6).

Figure 6. Inhomogeneous mixture of shredded textile waste.

3.2.2 Characterization of thermal insulation material matrix

Polystyrene waste can come from expanded polystyrene thermal insulation boards resulting from renovations, rehabilitations or demolitions of buildings, or from the collection and recycling of packaging used to protect equipment during transport.

Low density is characteristic of expanded polystyrene. A lower density contributes to ease of material handling and transport, but may affect mechanical strength. In general, low density is favorable for insulation applications because it reduces the overall weight of the construction. The low density and excellent thermal conductivity confirm the efficiency of expanded polystyrene as an insulation material.

3.2.3 Technological process of TermoText boards

The plates were obtained using the heat press shown in Figure 7, a specially designed equipment for conducting experiments within the Research Center of the Faculty of Industrial Design and Business Management, TUIASI.

The values indicate a low thermal conductivity, suggesting that the TermoTex material is effective in preventing heat loss. The coefficients obtained are competitive and suggest a performance similar to other insulating solutions available in the market and even better than polystyrene of 0.04 W/mK. By introducing textile waste into the mix, at the production temperature of 110°C, the thermal performance of the TermoTEX insulating panels has been improved.

(a) Settling the mixture (b) Thermopress and cooling (c) TermoTex panel

Figure 7. The process of making the plates in the thermal press.

15% EPS 25% EPS 20% EPS

Figure 8. TermoTEX boards made at 110°C.

The different variants of the manufactured boards are illustrated in Figure 8. The thermal insulation boards that presented the lowest values for all measured parameters were produced under experimental conditions using the following technological settings: 25% expanded polystyrene and a heat treatment duration of 15 minutes at a temperature of 110°C. The properties of the optimal TermoTEX variant are:

- Thermal conductivity coefficient (λ) = 0.033 W/mK
- Hygroscopicity (H) = 3.8%
- Density (ρ) = 158 kg/m^3

The values indicate a low thermal conductivity, suggesting that the TermoTex material is effective in preventing heat loss. The coefficients obtained are competitive and suggest a similar performance to other insulation solutions available in the market and even better than polystyrene (0.04 W/mK).

3.2.4 Characterisation of the composite material

The thermal characteristics study was carried out in the laboratories of the Faculty of Construction and Installations in Iaşi, in the double climate chamber, manufactured by Freutron Klimasimulation GmbH from Germany. This test apparatus offers the possibility to create two distinct environments in the two independent chambers, each with its own temperature and humidity characteristics. The cold room can simulate temperature conditions between –45°C and +100°C, with humidity ranging from 15% to 95%, while the warm room allows the creation of an environment with temperatures between +5°C and +100°C, and humidity between 10% and 95%. The results were processed with the PC208W software.

Hygroscopicity represents the ability of a material to absorb moisture from the environment, an essential phenomenon in the evaluation of the performance of construction materials, especially in the context of thermal insulation. This property influences not only the interior comfort of buildings, but also their durability and energy efficiency. The testing was carried out with the help of the climatic chamber presented in the previous subchapter. The test temperature and the percentage of relative humidity were chosen from the average of minimum temperatures to an average of maximum humidity recorded in Romania in 2022 according to the annual report made by the National Meteorological Administration, respectively a temperature of 13°C and a humidity of 71%.

Highlighting the influence of TermoTEX panels used in thermal insulation systems for the insulation of the external walls of buildings, was done by using specialized programs for the design and analysis of assemblies. The programs used were: Autocad (Grosu, 2015), which provides graphic models and RDM 6.17, which provides the thermal indicators, the studied parameters being the analysis of the heat flows of each type of wall analyzed and their thermal resistance.

The characteristic elements of each type of flow are the following: the internal temperature ($T_i = 20°C$) and the external one ($T_e = 18°C$), B – the width of the domain for each analyzed wall, where $B \leq 1.2$ m, and in the numerical analyzes considered $B = 1.2\ m$, as well as the linear thermal transfer coefficient (ψ), (Baciu et al., 2020).

The linear heat transfer coefficient„ represents the excess flow ($\Delta\Phi$) transmitted through a linear thermal bridge, related to its length (ℓ) and to the total temperature drop (ΔT), which represents the difference between indoor and outdoor air temperatures (Ştefănescu, 2012).

After the realization of the structural configurations with the specific details, for each case study, the results of the numerical analysis performed with the RDM 6.17 software application, in accordance with the regulations in force, are presented. Based on the determined values, observations can be made regarding the influence of the insulating panels on the thermal resistance of the walls and the improvement of the thermal insulation through the use of TermoTEX panels, highlighted graphically in Figure 9.

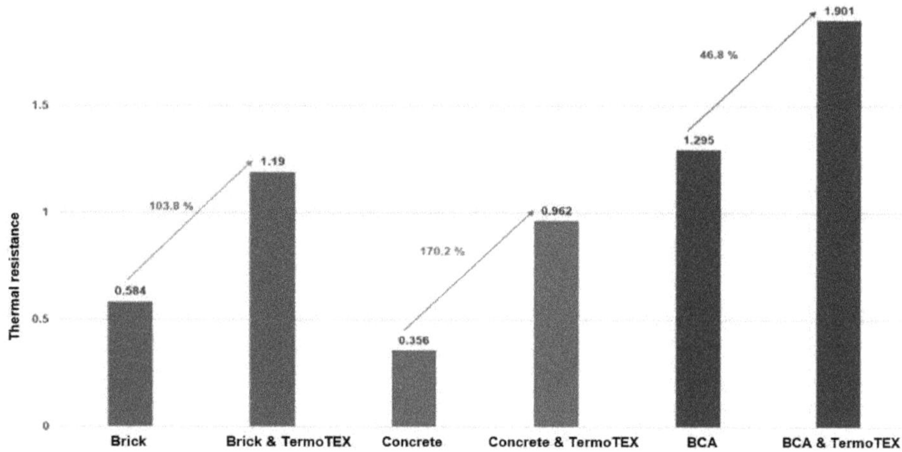

Figure 9. Comparison of linear heat transfer coefficients of brick, concrete and BCA structures, with and without TermoTEX layer.

Thus, from Figure 9 it can be seen:

- **Increase in thermal resistance**: In all three types of walls analyzed (brick, reinforced concrete, BCA), the addition of TermoTEX panels leads to a significant increase in the thermal resistance of the wall. This means that walls insulated with TermoTEX have a much greater ability to retain heat inside the building, which leads to a reduction in energy requirements for heating.
- **Energy efficiency**: TermoTEX panels improve the thermal performance of all types of walls, but the strongest impact is observed in reinforced concrete walls, which have a low thermal resistance, and after installing the TermoTEX panel with the thickness of 2 cm, the thermal resistance increased by 170.2%.
- **Thermal insulation performance**: Even though BCA already has a high thermal resistance, the TermoTEX layer further improves the insulation, which makes this type of wall present a superior thermal insulation performance, in comparison even with traditional structural insulation materials.

The comparison of the linear heat transfer coefficients of brick, concrete and BCA structures, with and without the TermoTEX layer, unequivocally emphasizes the contribution brought by the TermoTEX to improve the thermal performance of the external wall structures analyzed in the three case studies.

4. Conclusions

Any product, process or service can be evaluated through the Life Cycle Assessment (LCA) technique to quantify the impact on the environment during its life cycle (ISO-14040, 2006). With the help of LCA studies, we have the opportunity to propose, develop and implement various solutions to optimize the related phases of the product or service life cycle. These studies are the most powerful tools for sustainability analysis and interpretation.

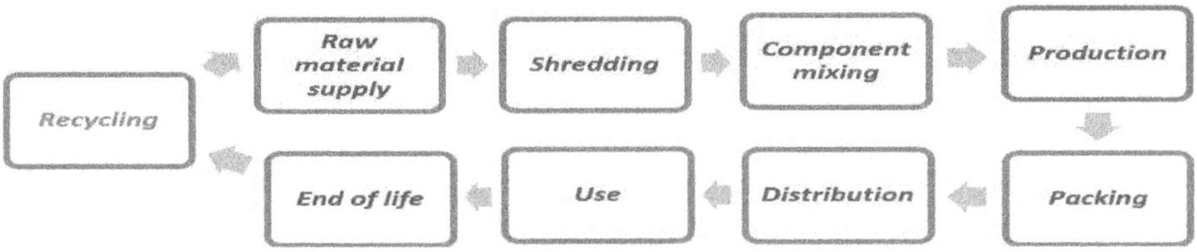

Figure 10. Life cycle stages.

The quantification of the impact on the environment through the life cycle assessment of the proposed innovative solutions has an essential role in assessing their ecological viability. The most accurate scenario in these two cases is the *closed-loop* scenario which allows the calculation of the environmental impact delimited from *cardle to cardle*. Thus, the overview of the entire product life cycle, including the end-of-life recycling scenario, presented in Figure 10, will provide valuable data for improving environmental performance from the early stages of innovation.

Carrying out the life cycle analysis for the TermoTEX thermal insulation panels, the ReCiPe 2016 Endpoint v1.08 method was used, which includes global normalization reference characterization factors from 2010. The cradle-to-gate evaluation system starts from the raw material, in this case the waste textiles and polystyrene, and analyzes the impact throughout the production of the product, until it leaves the factory gate,

The analysis takes into account the energy used to process the raw materials and manufacture the thermal insulation panels and includes end-of-life product management scenarios.

The comparative analysis based on the IPCC method revealed that TermoTEX boards have a lower global warming potential (GWP) compared to expanded polystyrene (EPS). This result indicates that TermoTEX contributes less to climate change in the long term, offering a more favorable option in the context of global warming.

In the evaluation carried out with the ReCiPe method, TermoTEX demonstrated a smaller footprint on the decrease of fossil resources compared to EPS. This emphasizes the greater efficiency of TermoTEX tiles in the use and conservation of non-renewable resources, indicating a more sustainable approach to the use of natural resources.

The quantification of the environmental impact through a cradle-to-cradle life cycle assessment of the composite panels FRC RECOTEX was carried out using the GaBi software. Since the proposed composite material is intended as an alternative to OSB type composite panels, the results obtained will be compared to establish what impact this would have on the environment.

Considering the values obtained over the entire LCA/Life Cycle Analysis, it can be stated that the composite material has an improved impact on the environment, compared to the OSB, which has the lowest impact in 11 categories.

However, we cannot fail to note the benefits of the composite material, which registers lower values of the impact parameters (human toxicity, carcinogenic effects and eutrophication), which indicates that it has less harmful effects on human health and biodiversity compared to OSB boards.

At the same time, we note that the analysis made on the production stage of the composite material recorded better values than the OSB, a fact that confirms to us that the proposed solution for waste recycling is a solution that would bring important environmental benefits.

Although the composite material made of polyester and polypropylene waste has an almost double density (1112 kg/m³) compared to a wood-based product, the performance results are still satisfactory, even under these less favorable conditions. It is well known that wood during the vegetative period stores carbon dioxide and releases oxygen, being considered a negative carbon material, a fact that was evident in the values obtained by OSB.

The impact of disposal to landfills of textile waste at the end of the life cycle is considered identical for all types of fiber (700 CO_2/kg) which confirms the positive advantage of recycling on the environment (Potting and Blok, 1995; Sandin and Peters, 2018; Thompson et al., 2012). The proposed composite material addresses the growing issue of textile waste accumulation by contributing to landfill decongestion and mitigating environmental impact. Moreover, its integration into the construction sector, where the demand for building materials is steadily increasing, offers a viable alternative to wood-based panels, thereby helping to reduce the consumption of natural timber resources. Thus textile waste can replace the need for wood raw material with beneficial consequences for the biosphere.

In conclusion, the two proposed innovative materials can contribute to reducing the negative ecological impact generated by the growing flow of textile waste by decongesting landfills and capitalizing on an important source of raw material by transforming and transferring it to another field where the demand for construction materials is constantly growing. Thus, boards made from textile waste can replace part of the raw material needed in the production of OSB boards and thermal insulation boards, with beneficial consequences for the biosphere.

References

Arena, U. (2012). Process and technological aspects of municipal solid waste gasification. A review. Waste Management 32(4): 625–639. doi:10.1016/j.wasman.2011.09.025.

Avadanei, M., Olaru, S., Ionescu, I., Florea, A., Curteza, A., Loghin, E. et al. (2021). Clothing development process towards a circular model. Industria Textilă. 72: 89–96.

Baciu, I.-R., Isopescu, D. N., Taranu, N., Lupu, M. L. and Maxineasa, S. G. (2020). Comparative analysis of the effect of different types of green roofs over the linear thermal bridges. In IOP Conference Series: Materials Science and Engineering (Vol. 916, Issue 1, p. 012004). IOP Publishing. https://doi.org/10.1088/1757-899x/916/1/012004.

Bell, N. C., Lee, P., Riley, K. and Slater, S. (2018). Tackling problematic textile waste streams. RESYNTEX. Available online: http://www. resyntex. eu (accessed 26 July 2018).

Ben-Eli, M. U. (2018). Sustainability: Definition and five core principles, a systems perspective. Sustain. Sci. 13: 1337–1343; 4.

Beton, A., Dias, D., Farrant, L., Gibon, T. and Le Guern, T. (2019). Environmental Improvement Potential of textiles (IMPRO Textiles). Publ. Off. Eur. Union 2014, 9. Online: https://ec.europa.eu/jrc/en/

publication/eur-scientific-and-technical-research-reports/environmental-improvement-potential-textiles-impro-textiles (accessed 5 March 2020).

Bhuiyan, M. A. R., Rahman, M. M., Shaid, A., Bashar, M. M. and Khan, M. A. (2015). Scope of reusing and recycling the textile wastewater after treatment with gamma radiation. Journal of Cleaner Production 112: 3063–3071. https://doi.org/10.1016/j.jclepro.2015.10.029.

Bodoga, A., Nistorac, A., Dragomir, A., Ailenei, E. C., Seul, A., Diaconu, M., Balan, C. D. et al. (2023). Ozone–vacuum-based decontamination: balancing environmental responsibility and textile waste. Sustainability 15: 16068. https://doi.org/10.3390/su152216068.

Boone, T. (2009). Sustainable Fashion and Textiles: Design Journeys by Kate Fletcher (Earthscan, 2008), Fashion Practice, 1(2): 271–274, DOI: 10.2752/175693809X469229.

Briga-Sá, A., Nascimento, D., Teixeira, N., Pinto, J., Caldeira, F., Varum, H. et al. (2013). Textile waste as an alternative thermal insulation building material solution. Construction and Building Materials 38: 155–160. doi:10.1016/j.conbuildmat.2012.08.037.

Chemicals in textiles -Risks to human health and the environment Report from a government assignment REPORT 6/14. (n.d.). https://www.kemi.se/download/18.6df1d3df171c243fb23a98f3/1591454110491/rapport-6-14-chemicals-in-textiles.pdf.

Conke, L. S. (2018). Barriers to waste recycling development: Evidence from brazil. Resources, Conservation and Recycling 134: 129–135. doi:10.1016/j.resconrec.2018.03.007.

Ekström, K. M. and Salomonson, N. (2014). Reuse and recycling of clothing and Textiles—A network approach. Journal of Macromarketing 34(3): 383–399. doi:10.1177/0276146714529658.

Elander, M. and Ljungkvist, H. (2016). Critical aspects in design for fiber-to-fiber recycling of textiles. A Mistra Future Fashion Report Phase, 2. Accessed at http://mistrafuturefashion.com/wp-content/uploads/2016/06/MFF-report-2016-1-Critical-aspects.pdf.

Environmental Protection Agency (EPA). (2021). Plastic Waste Management. Disponibil la: epa.gov.

European Commission (EC). (2022). Analysis of the national long-term renovation strategies. pp. 10–155. Retrieved from https://energy.ec.europa.eu/system/files/2022-12/SWD-Analysis-of-2020-LTRS.PDF.

Global Fashion Agenda and The Boston Consulting Group. (2017). Pulse of the fashion industry. Authors: Jonas Eder-Hansen, Caroline Chalmer, Sofia Tärneberg, Thomas Tochtermann, Retrieved from https://globalfashionagenda.com/?x=0&y=0&s=Pulse-of-the-Fashion-Industry_2017.

Grosu, M.-C. (2015). Modelling and simulation of heat transfer through Țurcana wool insulated walls. Industria Textilă 66(1): 48–54.

Hawley, J. M. (2006). Textile recycling: A systems perspective. In Recycling in Textiles. Woodhead Publishing Limited, UK.

ISO 14044:2006, Environmental management – Life cycle assessment – Requirements and guidelines.

KEMI. (2014). Report 6/14: Chemicals in textiles-Risks to human health and the environment Report from a government assignment. Publisher: Kemikalieinspektionen. ISSN:0284-1185. https://www.kemi.se/en/publications/reports/2014/report-6-14-chemicals-in-textiles.

Koenhemsi, D. (2017). Price alert: Are lumber prices rising too fast? Think again with the OSB in sight. Online: https://ihsmarkit.com/research-analysis/price-alert-are-lumber-prices-rising-too-fast-think-again-with-the-osb-in-sight.html.

Latan, H., Chiappetta Jabbour, C. J., Lopes de Sousa Jabbour, A. B., Wamba, S. F. and Shahbaz, M. (2018). Effects of environmental strategy, environmental uncertainty and top management's commitment on corporate environmental performance: The role of environmental management accounting. Journal of Cleaner Production. https://doi.org/10.1016/j.jclepro.2018.01.106.

Makrygiannis, I. and Karalis, K. (2023). Optimizing building thermal insulation: The impact of brick geometry and thermal coefficient on energy efficiency and comfort. Materials Science. Free Online Library.

Marconi, M., Landi, D., Meo, I. and Germani, M. (2018). Reuse of tires textile fibers in plastic compounds: Is this scenario environmentally sustainable? Paper presented at the Procedia CIRP 69: 944–949. doi:10.1016/j.procir.2017.11.074.

McGregor, L. (2015). Are closed-loop textiles the future of fashion?. Sourcing Journal. Accessed at: https://sourcingjournal. com/topics/raw-materials/are-closed-loop-textiles-the-future of-fashion-36800.

Muthu, S. S., Li, Y., Hu, J. Y. and Ze, L. (2012a). Carbon footprint reduction in the textile process chain: Recycling of textile materials. Fibers and Polymers 13(8): 1065–1070. doi:10.1007/s12221-012-1065-0.

Muthu, S. S., Li, Y., Hu, J. and Mok, P. Y. (2012b). Recyclability Potential Index (RPI): The concept and quantification of RPI for textile fibres. Ecological Indicators 18: 58–62. https://doi.org/10.1016/j.ecolind.2011.10.003.

Nistorac, A., Bodoga, A., Ailenei, C. and Loghin, C. (2024), Sustainable horizons: exploring technical textiles and environmental responsibility. International Symposium Technical Textiles - Present and Future, Edition 2023, Sciendo 2024, pp. 136–140.

Nixon, H. and Saphores, J.-M. (2009). Information and the decision to recycle: Results from a survey of US households. Journal of Environmental Planning and Management 52(2): 257–277. doi:10.1080/09640560802666610.

Nørup, N., Pihl, K., Damgaard, A. and Scheutz, C. (2018). Development and testing of a sorting and quality assessment method for textile waste. Waste Management 79: 8–21. doi:10.1016/j.wasman.2018.07.008.

Nørup, N., Pihl, K., Damgaard, A. and Scheutz, C. (2018). Development and testing of a sorting and quality assessment method for textile waste. Waste Management 79: 8–21. doi:10.1016/j.wasman.2018.07.008.

Oecotextiles. (2016). Why Do We Offer Safe Fabrics? Online document: https://oecotextiles.wordpress.com/category/co2-emissions-in-textile-industry/.

Opperskalski, S., Evonne, S. and Truscott, T. (2019). Textile-Exchange_Preferred-Fiber-Material-Market-Report_201. Text. Exch. 2019, 6. Accessed online: https://textileexchange.org/wp-content/uploads/2019/11/Textile-Exchange_Preferred-Fiber-Material-Market-Report_2019.pdf (accessed June 2020).

ONU. (2024). World Population Prospect 2024 ONU. Accessed online: https://www.un.org/development/desa/pd/world-population-prospects-2024.

Ouyang, P. and Cao, Z. (2012). Current status, problems, and path selection for recycling system construction in China. Recycle. Resour. Circ. Econ. 5(7): 21–24.

Papadaki, D., Nikolaou, D. A. and Assimakopoulos, M. N. (2022). Circular environmental impact of recycled building materials and residential renewable energy. Sustainability 14(7): 4039. https://doi.org/10.3390/su14074039.

Payne, A. (2015). Open-and closed-loop recycling of textile and apparel products. pp. 103–123. In Handbook of Life Cycle Assessment (LCA) of Textiles and Clothing. Woodhead Publishing. 20.

Plastics Europe. (2021). Plastics – the Facts 2021: An analysis of European plastics production, demand and waste data. https://plasticseurope.org/knowledge-hub/plastics-the-facts-2021/.

Potting, J. and Blok, K. (1995). Life-cycle assessment of four types of floor covering. Journal of Cleaner Production 3(4): 201–213. doi:10.1016/0959-6526(95)00082-8.

Rochman, C. M., Tahir, A., Williams, S. L., Baxa, D. V., Lam, R., Miller, J. T. et al. (2015). Anthropogenic debris in seafood: Plastic debris and fibers from textiles in fish and bivalves sold for human consumption. Scientific Reports 5 doi:10.1038/srep14340.

Sandin, G. and Peters, G. M. (2018). Environmental impact of textile reuse and recycling—A review. Journal of Cleaner Production 184: 353–365. doi:10.1016/j.jclepro.2018.02.266.

Sandin, Gustav and Peters, Greg. (2018). Environmental impact of textile reuse and recycling—A review. Journal of Cleaner Production. https://doi.org/10.1016/j.jclepro.2018.02.266.

SEPA. (2015). Circular Economy–elements for the new proposal. Accessed at: http://www.naturvardsverket.se/upload/miljoarbete-i-samhallet/miljoarbete-i-eu/cirkular-ekonomi/sveriges-standpunkt-eu-kommissionen-cirkul%C3%A4r%20ekonomi-150626.pdf.

SR EN. (2006). Oriented strand boards (OSB). Definitions, classification and specifications (Standard Roman- SR EN 300: 2006/SR EN 310), European Committee for Standardization, Brussels, 2006.

SR EN. (1996). Plăci de aşchii şi plăci de fibre. Determinare a umflării în grosime după imersie în apă. in Romanian (Standard Roman - SR EN ISO- 317: 1996), Romanian Standardization Association, Bucuresti, 2000.

Ştefănescu, D. (2012). Manual de proiectare higrotermică a clădirilor, Editura Societății Academice Matei-Teiu Botez, ISBN 978-606- 582-015-9.

Superti, V., Forman, T. V. and Houmani, C. (2021). Recycling thermal insulation materials: a case study on more circular management of expanded polystyrene and stonewool in Switzerland and research agenda. Resources 10(10): 104. https://doi.org/10.3390/resources10100104.

Thompson, P., Willis, P. and Morley, N. (2012). A review of commercial textile fiber recycling technologies. WRAP.[pdf] Retrieved March 9, 2017. Available at: http://www.wrap.org.uk/content/review-commercial-textile-fibre-recycling-technologies.

Truscott, L., Tan, E. and Opperskalski, S. (2017). Preferred Fibre & Materials Market Report 2018. Textile Exchange 1–96.

Textile Exchange. (2022). Preferred Fiber & Materials Market Report. Retrieved from https://textileexchange.org/knowledge-center/reports/2017-preferred-fiber-materials-market-report/.

United States Environmental Protection Agency. (2021). Trash-Free Waters Program: Plastics in the aquatic environment. U.S. EPA. https://www.epa.gov/trash-free-waters/plastics-aquatic-environment.

Chapter 10

University Sustainability

AI, Drivers, and Perspectives

Delia Rozovlean

1. Introduction

For a good understanding of all that sustainability means, we can consider the definition provided by the United Nations: "development that meets the needs of the present without compromising the ability of future generations to meet their own needs" (Commission of Nations United for the Environment and Development, 2002).

Sustainability is the concept that designates the ability to meet the needs of the present without compromising the chance of future generations to meet their own needs. It includes a balanced approach between economic, environmental and social aspects so as to encourage sustainable development (https://op.europa.eu/webpub/eac/education-and-training-monitor/en/; Rampelt et al., 2019).

Sustainable development has three major pillars: economic, social and environmental. Economic sustainability means creating an economic system that promotes prosperity and generates long-term benefits. This development is good to be cost effective and support communities. The second pillar of sustainability is the social one, closely related to the well-being of the population and the protection of the vulnerable. And the last pillar is the environmental one, where it is necessary to strike a balance between the resources consumed and those that the planet can generate. This refers to all types of resources, such as extensive raw materials or greenhouse gas emissions released into the atmosphere. Environmental sustainability

Politehnica University of Timisoara, Faculty of Management in Production and Transportation, 14 Remus Street, Timisoara, Romania.
Email: delia.rozovlean@upt.ro

DOI: 10.1201/9781003518990-10

requires good development focused on the use of renewable energies as well as low environmental impact processes based on circularity that can help recycle, recover and reuse resources (CE website, Education and Training; Rampelt et al., 2019).

In 2015, 17 Sustainable Development Goals (SDGs), also known as the 2030 Agenda, were adopted. They represent the global plan to ensure prosperity, well-being and environmental protection until 2023 (Commission of Nations United for the Environment and Development, 2002).

The 17 objectives adopted are:

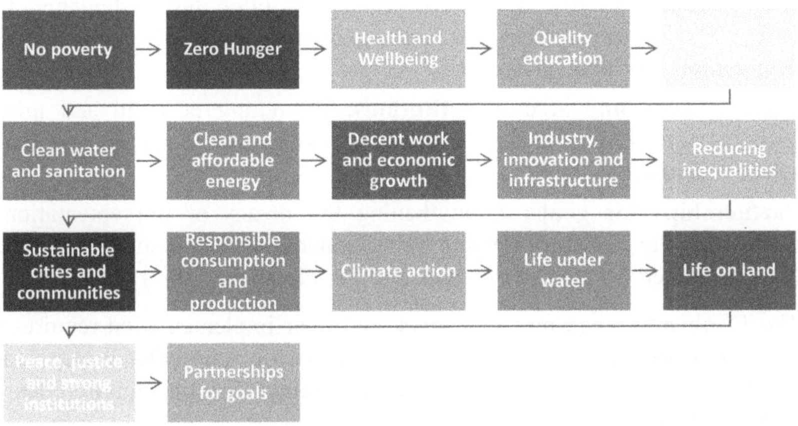

- **No poverty:** Eradication of poverty, in all its forms
- **Zero Hunger:** Eliminating hunger, improving nutrition and promoting sustainable agriculture
- **Health and Wellbeing:** Ensuring healthy living and promoting well-being for all ages
- **Quality education:** Ensuring quality, inclusive and equitable education and promoting lifelong learning opportunities
- **Gender equality:** Achieving gender equality and empowering women and girls
- **Clean water and sanitation:** Ensuring universal access to clean drinking water and sanitation
- **Clean and affordable energy:** Ensuring access to affordable, sustainable and modern energy for everyone
- **Decent work and economic growth:** Promoting sustained and sustainable economic growth, full and productive employment and decent conditions
- **Industry, innovation and infrastructure:** Building resilient infrastructure, promoting sustainable industrialization and encouraging innovation
- **Reducing inequalities:** Reducing inequalities between countries, but also within them

- **Sustainable cities and communities:** Developing safe, resilient and sustainable cities and human settlements
- **Responsible consumption and production:** Ensuring sustainable patterns of consumption and production
- **Climate action:** Taking urgent measures to combat climate change and its impact
- **Life under water:** Conservation and sustainable use of oceans, seas and marine resources
- **Life on land:** Protecting, restoring and promoting the sustainable use of terrestrial ecosystems, managing forests, combating desertification, halting soil degradation and biodiversity loss
- **Peace, justice and strong institutions:** Promoting peaceful and inclusive societies for sustainable development, access to justice for all and building effective and accountable institutions
- **Partnerships for goals:** Strengthening the means of implementation and revitalizing the global partnership for sustainable development (Commission of Nations United for the Environment and Development, 2002)

The 17 objectives are interdependent, and their implementation requires good collaboration between governments, the private sector, international organizations and civil society (Commission of Nations United for the Environment and Development, 2002).

Artificial intelligence has an important impact on university sustainability, contributing to an accelerated transformation through administrative and operational optimization through the improvement of the learning process, consistent support in research as well as the involvement of the university community through collaboration platforms (Rajani, 2011).

Administrative and operational optimization is divided into two categories, and in the first category is resource management where artificial intelligence is used by monitoring the consumption of water, energy or other resources helping to make universities more efficient, and another element would be the flattening of campuses with the help of artificial intelligence, by creating more energy efficient spaces (Rajani, 2011).

In the second category, we can discuss the improvement of the learning process, through the personalization of education where artificial intelligence can adapt study materials to the individual requirements and needs of students, helping to make the educational process more efficient (Rajani, 2011).

2. Education for sustainable development

Education for sustainable development is an important concept as a response to contemporary global challenges such as climate change, depletion of natural resources and social inequality. This concept promotes an integrated approach to learning that supports the development of skills needed to build a more sustainable future. Through education for sustainable development, people can be equipped with the knowledge and skills to build a more resilient, just, and sustainable world. It will help shape capable citizens, responsible for making ethical decisions to support the well-being of future generations.

a. **Adapted curriculum** refers to the process of combining several fields of study, disciplines or skills into a coherent educational plan. This involves making connections between different subjects to provide students with more in-depth, interconnected and contextualised learning. Integration in the curriculum represents an innovative approach that brings more relevance to education, therefore we have several types of integration in the curriculum: thematic integration, which focuses on a common subject, the second type is interdisciplinary integration, with concepts and methods from several disciplines, the third is transdisciplinary integration, focusing on real problems, the fourth is vertical curricular integration where the emphasis is on creating a logical progression of skills and concepts between different levels of learning and the last one being the horizontal curricular integration where contents that are taught simultaneously in several subjects are connected.
b. **Interdisciplinary research programs** are complex initiatives that bring together expertise from various fields to address problems that cannot be adequately addressed by a single discipline. The programs aim to integrate methodologies, perspectives and knowledge from multiple disciplines to create effective and innovative solutions. They bring added value, generating a lot of knowledge and better integrated solutions that contribute to the progress of society. Therefore, there are some key features of these programs such as: collaboration between disciplines where experts from various fields are also involved such as natural sciences, social sciences, engineering, the arts or medicine. The second characteristic would be complex problems that are oriented towards challenges that require approaches, such as climate change, global health, artificial intelligence or sustainability. The third characteristic is innovation, which presents the connection between fields and generates new perspectives, but also innovative solutions, overcoming the conventional boundaries of knowledge. The last characteristic being diverse methodologies combining qualitative and quantitative methods, fundamental and applied research, advanced technology and critical analysis.

2.1 Involvement of the university community

University community involvement is a concept that refers to the active participation of all members of a university (teachers, students, administrative staff and other collaborators) in various activities that can influence both university life and the development of the local, national or global community. This can add a varied range of initiatives and projects aimed at educational as well as social, cultural or economic goals (Taucean et al., 2019).

The university community can be actively involved by participating in the educational process of both teachers and students in academic projects, seminars or research (Taucean et al., 2019).

a. **Awareness campaigns** involving the university community are important in promoting responsible actions, to support education, but also to create a healthier environment. Universities were and are centers of education, but also of social change, and the active involvement of students, professors and administrative staff can significantly contribute to improving conditions in the academic community. Within a university community, several awareness campaigns can be exemplified such as: **environmental protection campaigns**, to encourage university staff and students to contribute to protecting the environment, **campaigns to promote the balance between academic and personal life**, where one can learn how to better manage time and **campaigns to promote civic engagement and volunteerism** to help change the university community.
b. **Student-organized** initiatives in the field of education for sustainable development are important because they encourage social responsibility, ecological awareness and active participation in solving global challenges.

3. The role of artificial intelligence for sustainable universities

Artificial intelligence plays an important role in promoting sustainability in universities by supporting institutions in reducing their environmental impact by optimizing resources and focusing on student education in a sustainable way (Rajani, 2011).

Artificial intelligence can contribute to the transformation of universities into sustainable entities in various ways such as: optimization of energy consumption, waste management and sustainable transport (Rajani, 2011).

3.1 Sustainable campus management

Sustainable campus management is based on a set of practices, initiatives and policies designed to reduce negative environmental impacts, support the well-being and health of the university community and promote resource efficiency.

There are several universities in the world that have implemented successful sustainable programs such as: Stanford University (USA), where a water saving program is generated, Copenhagen University (Denmark), is a leader in using renewable energy and driving CO_2 emissions and Babeş-Bolyai University (Ro), where ecological education is promoted and tree planting projects are organized (sursa:Web of Science - Clarivite) (Figure 1).

In the diagram below, the data of the 3 universities were analyzed regarding the sustainable management of the campuses and the pursuit of the Sustainable Development Goals (SDGs), providing opportunities in teaching and research (Commission of Nations United for the Environment and Development, 2002).

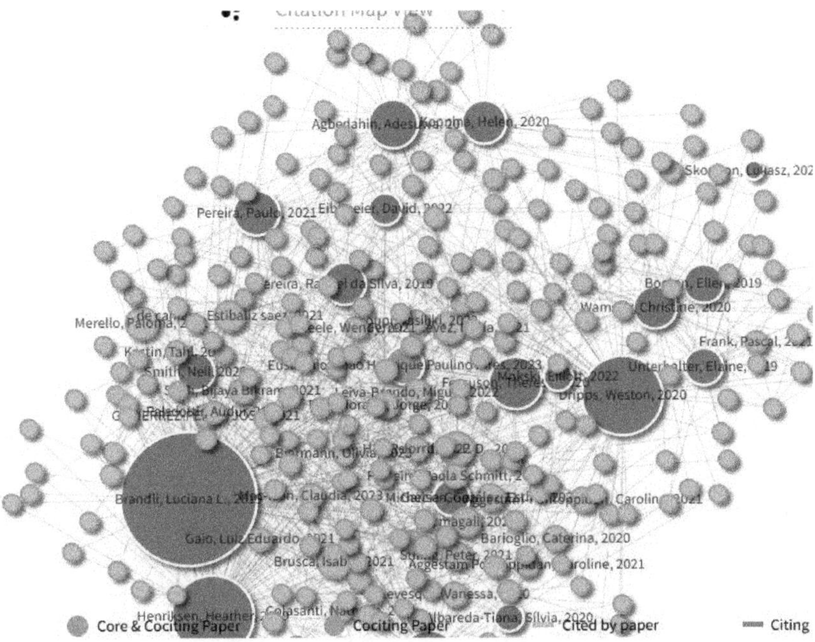

Figure 1. Grouping of co-authors from the 3 university centers. Source: Web of Science - Clarivite.

3.1.1 Reducing energy consumption

Reducing energy consumption in universities is essential and can be done by reducing costs and protecting the environment. This goal can be achieved through a number of behavioral and technological measures. There are a number of technological measures such as: energy-efficient lighting by replacing LED bulbs or installing motion sensors in laboratories or classrooms. In addition to technological measures, we can also discuss behavioral measures such as education and awareness by organizing awareness campaigns, informing students and staff about the importance of saving energy or promoting good practices on university campuses by creating volunteer teams.

3.1.2 Waste management

Waste management in universities is a primary aspect of campus sustainability and environmental education. The separate collection of waste through marked collection points, clear instructions and recycling stations in strategic areas such as classrooms, student dormitories, or libraries can contribute to the protection of the environment and the education of future generations in the spirit of ecological responsibility.

3.1.3 Sustainable transport

Artificial intelligence is a scientific field with great potential to create a technological and cognitive type of development and design of smart cities. In this study we examined the current role of artificial intelligence in transportation. With the help

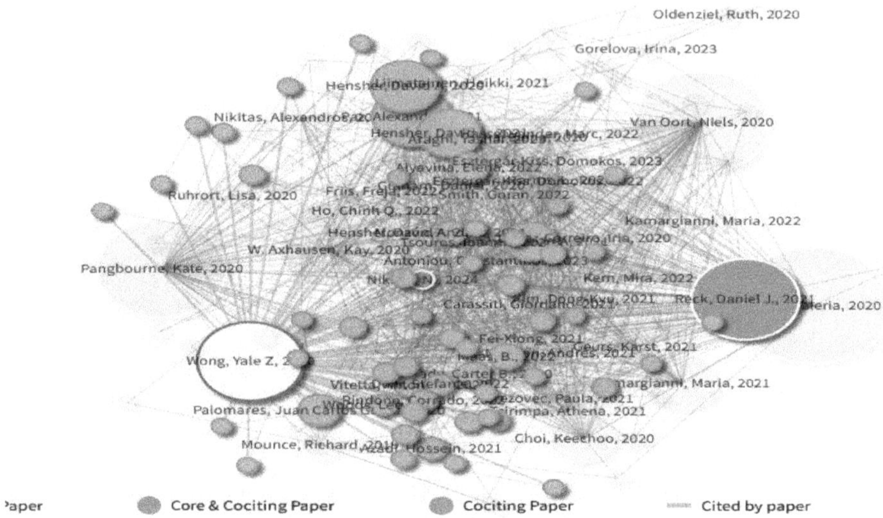

Figure 2. Cociting paper. Source: Web of Science - Clarivite.

of AI, optimal routes can be planned for private or public transport in university campuses, reducing emissions generated by vehicles (Figure 2).

In the diagram below, with the help of the Clarivate website, data was analyzed between the years 2020–2024, regarding sustainable transport where it is desired to change the mobility models regarding the examination of the information of two universities such as the Politehnica University of Milan and the Politehnica University of Madrid (sursa:Web of Science - Clarivite).

3.2 Sustainability research

Research in the field of sustainability represents a complex sector for the university environment, having an impact on education. The research aimed to respond to global challenges such as climate change, diminishing natural resources through sustainable solutions. Universities have an important role in sustainable research by training specialists capable of implementing sustainable solutions in various fields, but also by creating partnerships between universities and the private sector (Cassingena-Harper, 2003).

3.2.1 Renewable energy projects

Renewable energy projects implemented in universities are valuable projects that promote sustainability, reduce carbon footprints and help educate students about green energies. There are some important projects that can be implemented such as the installation of photovoltaic solar panels to generate electricity on campuses, where energy bills or CO_2 emissions are reduced. Another example would be wind turbines through which one can benefit from the continuous production of green energy or energy efficiency projects (modernization of buildings through thermal insulation and the implementation of LED lighting) (Cassingena-Harper, 2003).

3.2.2 Climate impact studies

Climate impact studies are complex analyzes that estimate the effects of climate change on the environment, society, economy and infrastructure. Studies are important to understand how climate phenomena influence different fields of activity.

These studies are based on several objectives, namely: the identification of vulnerabilities, risk assessment, strategic planning and the promotion of sustainability.

The University of London, Harvard University and the University of Oxford conducted climate impact studies on air transport demand in 2020. The results suggest that the share of the population that traveled by air between 2018–2023 was 11%. Data shows a small share of air travel is responsible for much of global warming (sursa:Web of Science - Clarivite).

Climate impact studies contribute to public awareness, as well as decision-makers' awareness of the effects of the environment (Figure 3).

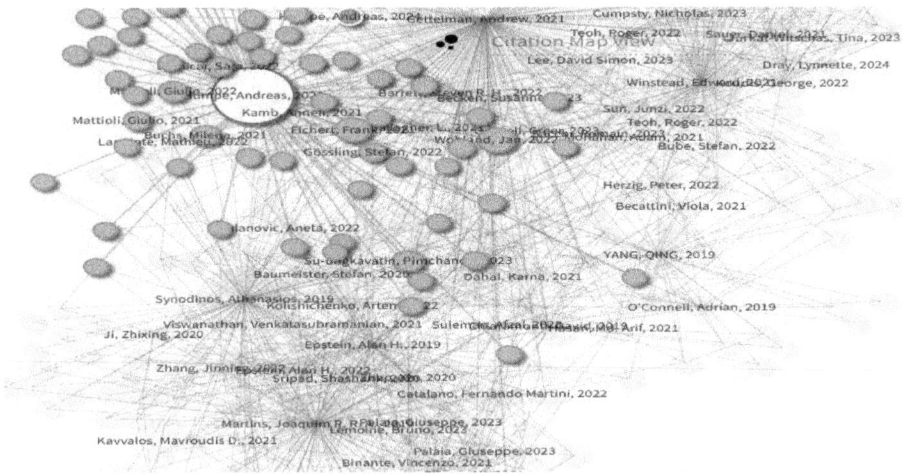

Figure 3. Research involvement of co-authors between 2018–2023. Source: Web of Science - Clarivite.

4. Insights into university sustainability and the future role of artificial intelligence

University sustainability means the ability of higher education institutions to implement, create and maintain long-term sustainable practices in education, research, community involvement and resource management.

University sustainability in the university environment is represented by: energy efficiency and the use of resources by implementing green technologies, education for sustainability by training future leaders prepared to tackle complex global problems, research oriented towards sustainable solutions where universities can be centers of innovation of problems related to changes climatic (Howe, 1994).

Artificial intelligence has an important impact on university sustainability, contributing to an accelerated transformation through administrative and operational optimization through the improvement of the learning process, consistent support in research as well as the involvement of the university community through collaboration platforms (Howe, 1994).

Administrative and operational optimization is divided into two categories, and in the first category is resource management where artificial intelligence is used by monitoring the consumption of water, energy or other resources helping to make universities more efficient, and another element would be the flattening of campuses with the help of artificial intelligence, by creating more energy efficient spaces (Howe, 1994).

In the second category, we can discuss the improvement of the learning process, through the personalization of education where artificial intelligence can adapt study materials to the individual requirements and needs of students, helping to make the educational process more efficient (Howe, 1994).

In addition to this process, basic content for artificial intelligence can also be created, and an example would be interactive stimulations helping students solve complex problems including those related to sustainability (Howe, 1994).

In conclusion, universities that adopt artificial intelligence and sustainability can become leaders in education for a sustainable future, having a positive impact both in the university environment and on communities.

5. Conclusions

The two big fields such as artificial intelligence and university sustainability represent fields that intersect more and more frequently, having the ability to contribute to the sustainable development of higher education institutions.

University sustainability refers to the integration of sustainable development principles in all sectors of university operation, namely campus administration, adapted curriculum, research promotion focused on innovative solutions for global problems, but also collaboration with other institutions.

Artificial intelligence supports universities in implementing sustainability initiatives by optimizing resources, research, education, global partnerships or smart campuses.

In the future, artificial intelligence and university sustainability may become interconnected. Universities will become centers of innovation in sustainability using artificial intelligence to shape and inspire positive change in society.

The two fields, together, can efficiently and responsibly address global challenges, contributing to a more sustainable future.

References

Ansoff, I. and McDonnell, E. (1990). Implanting Strategic Management (Second edition). New York: Englewood Cliffs: Prentice Hall International Inc.

Bratianu, C. and Vasilache, S. (2010). A factorial analysis of the managerial linear thinking model. International Journal of Innovation and Learning 8(4): 393–407.

Bartram, J., Brocklehurst, C., Bradley, D., Muller, M. and Evans, B. (2018). Policy review of the means of implementation targets and indicators for the sustainable development goal for water and sanitation. Clean Water 1(1): 3.

Berger, P. L. and Luckmann, T. (1967). The Social Construction of Reality, London: Penguin International, 9: 424–437.

Cassingena-Harper, J. (2003). Exploring Knowledge Futures in Information and Communications Technologies and Education in 2020 - Report on eFORESEE Malta Foresight Pilot. Kalkara: Malta Council for Science & Technology.

Cassingena-Harper, J. (2013). Impact of Technology Foresight. Compendium of Evidence on the Effectiveness of Innovation Policy Intervention, Manchester: Manchester Institute of Innovation Research, The University of Manchester.

Church, Chris. (1997). "Experienţe Utile în Dezvoltarea Durabilă", ed. Academia Română.

Commission européenne, Stratégie de l'Union européenne en faveur du développement durable, 2002.

Deloitte Insights. (2019). Tech Trends 2019. Beyond the Digital Frontier, London: Deloitte.

Dumitrescu, I., Iordache, O. G., Mitran, E. C., Stefanescu, D., Varzaru, E., Pislaru, M. et al. (2018). Multi-functional effects of textiles dyed with madder roots powder (Rubiatinctoria). Industria Textila 69(6): 451–457.

Global University Network for Innovation. (2019). Implementing the 2030 Agenda at Higher Education Institutions: Challenges and Responses, Barcelona: Global University Network for Innovation (GUNi).

Howe, J. (noiembrie 1994). Artificial Intelligence at Edinburgh University: A Perspective. Accesat în 30 august 2007.

Ivascu, L., Mocan, M., Draghici, A., Turi, A. and Rus, S. (2015). Modeling the green supply chain in the context of sustainable development. 4th World Conference on Business, Economics and Management (WCBEM-2015) 26: 702–708.

Ivascu, L. and Cioca, L. I. (2019). Occupational accidents assessment by field of activity and investigation model for prevention and control. Safety Journal 5(1): 12.

Rajani, Sandeep. (2011). Artificial intelligence – man or machine (PDF). International Journal of Information Technology and Knowledge Management 4(1): 173–176. Arhivat din original (PDF) la 18 ianuarie 2013.

Rampelt, F., Orr, D. and Knoth, A. (2019). Bologna Digital 2020. White Paper on Digitalisation in the European Higher Education Area, Berlin: Hochschulforum Digitalisierung (HFD).

Sarfraz, M., Ivascu, L., Belu, R. and Artene, A. (2021). Accentuating the interconnection between business sustainability and organizational performance in the context of the circular economy: The moderating role of organizational competitiveness. Business Strategy, and the Environment 30(4): 2108–2118.

Taucean, I., Tamasila, M., Ivascu, L., Miclea, S. and Negrut, M. (2019). Integrating sustainability and Lean: SLIM method and enterprise game proposed. Sustainability Journal 11(7): 2103.

Trandabat, A. F., Pislaru, M. and Brinzila, M. (2012). SiadEnv - Environmentally friendly e-building energy management system. Environmental Engineering and Management Journal 11(3): 687–694.

Zaman, G. and Gherasim, Z. (2006). Criteria, and principles of sustainable development from the point of view of its resources, Bulletin AGIR 4: 137–139.

Chapter 11

Sustainable Production in the Pharmaceutical Industry, Waste Management and Environmental Protection

Laura Crina Miraute (COCA),[1,*] *Ion Verzea*[2] and *Nicoleta Mihaela Căsăneanu (DASCĂLU)*[2]

1. Introduction

The interweaving of the three dimensions of sustainability—economic, social, and environmental—can transform the pharmaceutical industry into a fair and balanced one. Carrying out the activity of a company in a balanced way, both in terms of concern for employees, expenses/income, and resources consumed or concerns for the environment, could threaten less of future generations who will carry out their activity on this same planet with the same resources. Concerns about the planet that has been "hosting" us for many years are becoming increasingly prevalent today, particularly among the manufacturing industries that significantly impact the environment. Human actions impact everything related to life on this earth; therefore, the need for information and awareness is constantly increasing in support of future generations who do not have to suffer because of the present actions of humanity. The sustainability idea appeared many years ago when perhaps the situation was not as critical as it is today. Still, it was expected what the result would be if no

[1] "Gheorghe Asachi" Technical University of Iasi-Romania, Faculty of Industrial Design and Business Management, Prof. D. Mangeron Blvd., 700050, Iasi, Romania.
Emails: ion.verzea@academic.tuiasi.ro, nicoleta-mihaela.dascalu@student.tuiasi.ro
* Corresponding author: laura-crina.miraute@student.tuiasi.ro

DOI: 10.1201/9781003518990-11

action was taken at the right time. The Bruntdland Report (Brundtland, 1987) defines sustainability "as the ability of present generations to meet their needs without disturbing the ability of future generations to meet their own needs". After 1970, the specialized literature accepted two different groups of explanations of sustainability. One was linked to the notion of "sustainable development," and the other used a systemic perspective. There is no clear distinction between "durability" and "sustainability" in English, and generally the term "development" is used together with "evolution." Elkington first defined the concept of the triple baseline (TBL) to treat each of the three dimensions of sustainability in an equitable manner: social, economic, and environmental (Elkington et al., 1999). The sustainability approach encourages organizations around the world to adopt dynamic systems that support sustainable business performance. Implementing sustainable business practices has become a challenge that must be addressed quickly by adopting new tools (Bhatta et al., 2023).

Despite the known limits of capital and the world's absorptive capacities, international resource exploitation tripled between 1970 and 2017 and continues to increase globally (Shah et al., 2023).

Sustainability assessment is essential to an organization's development for several reasons: people are more receptive to environmental products; production contributes to the efficient increase of financial and environmental performance; corporate strategic decision-making must be based on a sustainability analysis, and local and national policies must integrate sustainability principles (Pislaru et al., 2019).

Based on what is found in the literature, among the three dimensions of sustainability, for companies in the pharmaceutical industry, the greatest concerns that exist are those for the environment. In pharmaceutical production companies, a very complex and rigorous process of manufacturing drugs takes place. From the early stage, it is based on an active substance, which is essential in the treatment of various diseases but is also the one that later, when it needs to be eliminated or destroyed, becomes dangerous for the environment.

One of the elements contributing to land contamination and the growing trend of environmental deterioration is the careless disposal of waste (Ofori, 2023). Interest in the circular economy has grown with the increasing awareness of environmental issues. Turning waste into resources is the typical definition of the circular economy (Shah et al., 2023). Consequently, experts, successive government organizations, scientists, and non-governmental organizations have initiated their projects on the measures that can be applied so that the results of their activity are sustainable. COVID-19 has helped nations increase the adoption of the circular economy, as it has shown the importance of reducing waste, showing that congestion could increase contamination and infections (Abbasi and Hassan, 2023).

The goal to lessen these perplexing human mistakes and to recover resources from waste instead of discarding them at great expense and environmental harm gave rise to the circular economy. This involves extending a product's life cycle while reducing its ecological footprint associated with production and distribution. Concerns about the changing climate, which are seen as a calamity of the commons, have also drawn attention from the rise in global temperatures and environmental degradation that contributes to it. Consequently, many nations have moved from economic growth to economic development, simultaneously with strong industrial expansion. This is vital because economic growth, progress, and development of production are linked to the generation of waste, especially emissions of methane, carbon dioxide, and carbon monoxide gases (Abbasi and Hassan, 2023). The idea of the circular economy originated from calls for a sustainable environment to lessen the environmental risk posed by waste creation. The circular economy's concept uses waste generation to generate profit-oriented activity while preserving the environment. As part of this, carbon dioxide and carbon monoxide emissions linked to careless waste disposal must be decreased (Ofori, 2021).

The circular economy has emerged as an innovative business theory with the potential to change the way companies engage in social responsibilities. Nowadays, it is a popular concept that aims for the future of sustainable progress and is promoted by the European Union and many other national governments.

The impact of the changes that may occur in the pharmaceutical industry and how these changes could improve people's quality of life and leave a better world for future generations depends on how much academic research contributes to this field. The well-defined and often short-term expiration date for pharmaceutical items, coupled with uncertainty regarding market demand and production rates, can lead to a substantial amount of pharmaceutical waste—that is, wasted or expired medicines that are no longer needed or usable. Due to its potential to have a negative impact on both the environment and human health, this form of waste has become an increasing problem (Silva et al., 2023).

Following this introductory section, there is a literature review, research methodology, results and discussions, and finally, the conclusions.

2. Literature review

The significance of supply chain sustainability has grown over the years as a result of growing awareness of the environmental and social effects of corporate operations. In addition, sustainability requirements in the modern business environment are increased by government forces and public demand. The pharmaceutical supply chain is one of the most important networks in the healthcare system. Managing and planning this chain at a strategic and operational level is essential for improving the health indicators of society. Validating that patients receive medicines at the right time and place is the main goal of the pharmaceutical supply chain (Farhad Safaei and Naeme Zarrinpoor, 2024). Because the environment is already under significant pressure from pollution, environmentalists are actively involved in various discussions about sustainability, a concept that has evolved over time and is currently controversial (Marius et al., 2017).

When designing an efficient pharmaceutical supply chain, many things need to be taken into account, such as the location of the facility, drug corruption, inventory control, financial approaches, suppliers, and end customers. It is worth noting that if the order does not arrive on time, there is an increased risk to patient health, as well as a higher probability of customer loss and higher costs for drug shortages. On the other hand, if a larger quantity of medicines is anticipated to be ordered than usual, there is a chance that the medicines will expire while in storage because most of them have limited expiration dates. Therefore, cost, quantity, and expiration date are crucial considerations in solutions for getting pharmaceuticals to the patient when they need them (Darestani and Hemmati, 2019).

Pharmaceutical manufacturing waste is unwanted residues or effluents containing active pharmaceutical ingredients and volatile organic substances. This waste originates from pharmaceutical manufacturing processes or the pharmaceutical industry and poses a significant hazard to the environment and health (Emmanuel Sunday Okeke et al., 2022).

According to the www.who.int/publications/i/item/9789241565585, 15% of pharmaceutical waste is toxic and 85% is non-toxic. Appropriate waste management methods are required for this type of waste. As it contributes significantly to the health of society, the industry is most responsible for implementing sustainable practices. The pharmaceutical industry pays attention to "access to medicines"—that is, ensuring that everyone, especially the poor in third world nations, have access to generic medicines—this is considered a major sustainability concern, as part of the pharmaceutical industry's commitment to human rights. Another significant issue facing the pharmaceutical sector is "hazardous waste". This is because the production of pharmaceuticals produces a large amount of hazardous waste (Wan, 2014).

Concerns about the harmful effects on wildlife caused by pharmaceutical residues have increased in the past few years (Brodin et al., 2017). A significant number of pharmaceutical products pollute surface water around the world, according to studies in the specialized literature (Wilkinson et al., 2022). Human excretion of medicines and their metabolites into the sewage systems is the primary source of the majority of pharmaceutical residues found in the environment. Densely populated locations or underdeveloped nations with poor sanitation typically have the biggest levels of contaminants (Wilkinson et al., 2022). Some nations have modified their methods for classifying environmental risks to take into account the hazards associated with the use of medicines. This was done by the regulatory guidance on environmental risk assessment (Sanja Riikonen et al., 2024). In addition, one of the main causes of pharmaceutical contamination in some places is the manufacturing of pharmaceuticals where there are insufficient environmental laws to control the licensing of pharmaceutical factories. As the production of pharmaceutical raw materials has been concentrated in regions with limited resources, alarming levels of contamination with pharmaceuticals have resulted from a lack of control, particularly in Asia (Larsson, 2014). In addition to the exposure of pharmaceutical waste to the environment, the production of medicines (including the consumption of water and clean energy), distribution (including storage and transportation), and end-of-life (including unused or thrown-away medications) all have an impact on the environment.

One recent study that comes to mind is the strategy put forth by Goodarzian et al. (2021). According to these writers, pharmaceutical supply chain administration must balance cost reduction and service standards while accounting for the risks associated with unpredictability. The supply chain system for pharmaceuticals, which includes production, distribution, procurement, ordering, holding, inventory, allocation, and routing under uncertain circumstances, is described in this study. There are several tiers, goals, and periods in the network. From a methodological perspective, a robust fuzzy programming strategy was proposed to address the uncertainty parameters in a mixed-integer nonlinear programming model. According to (Ahmadi et al., 2018), the pharmaceutical supply chain faces difficulties at three levels of decision-making: tactical, strategic and operational. They integrated the method of optimization into the management of the pharmaceutical supply chain. A scheme of good management practices has become an integrated part of the pharmaceutical supply chain. A real case study is described, and analyzed, and finally, a series of additional investigation methods are proposed. By introducing the most relevant elements and a problem-oriented modeling method, all the studies mentioned above have their limitations but can be overcome.

Using distributed ledger technology, Hulea et al. (2018) offered a solution for the pharmaceutical management of cold chains and suggested an ideal distribution of medicinal products. Additionally, they suggested cargo monitoring software to inform all parties involved. They employed the Hyperledger Sawtooth distributed ledger system to gather data. The suggested study effectively conveyed the system entities and modeled the data (Hulea et al., 2018). Generally speaking, prior research has given insight into the present issues and several potential solutions required to enhance the ecological responsibility of pharmaceutical companies. However, prior research has concentrated on certain development challenges, including more environmental friendly pharmaceutical goods design, cleaner production, or healthier supply chains. Usually, this is accomplished by looking at data that has been made public by businesses, gathered through surveys, or conducted through organized interviews.

On the other hand, the circular economy is a concept that can contribute to supporting sustainability in pharmaceutical companies. The circular economy can have a positive impact on people by creating jobs and profitable business models. Studies show that the circular economy could reduce poverty when properly implemented. According to one study, the circular economy can improve people's livelihoods in developing and emerging countries. Also, a study by Fernandes (2016) in Brazil found that circular methods have environmental benefits; there are opportunities to value people at the bottom of the pyramid; production and engagement can increase; and management has a responsibility to initiate an atmosphere of precise help. The circular economy can solve other developing economic problems, such as waste disposal, reducing poverty, using renewable energy, and creating more job possibilities (Gower et al., 2016).

3. Methodology

This research is a literature review, more precisely on articles that address the subject of interest, sustainability in the pharmaceutical industry. Utilizing the "Harzing Publish or Perish" software application, articles addressing the topic of interest in this research were extracted from the "Google Scholar" database, based on the keywords: "sustainability in the pharmaceutical industry" and "sustainability assessment in the pharmaceutical companies". The number of selected articles was limited to 200 for 8 years, from 2016 to 2024. In-depth research was done on the concept's approach utilizing the VOSviewer tool, which can categorize data taken from the "Google Scholar" platform based on the number of citations and keywords in the subject of sustainability evaluation of the pharmaceutical business.

Utilizing the "Harzing Publish or Perish" software application, articles addressing the topic of interest in this research were extracted from the Google Scholar database, based on the keywords: sustainability in the pharmaceutical industry and sustainability assessment in the pharmaceutical industry. The number of selected articles was limited to 200 for 8 years, from 2016 to 2024. Using the VOSviewer tool, which can classify details taken from the "Google Scholar" platform based to the number of citations and keywords in the field of sustainability assessment in the pharmaceutical industry, detailed research was conducted on the approach to the concept.

According to the results obtained in Table 1, the top publications are presented in the period 2016–2024. A peak of published works, according to Harzing or Publish, is in 2020, when 36 works are published, which represents 18% of the total of 200 highlighted articles. In 2021 and 2019, 31 articles were published, which means 15.5% for each year. In 2018, 29 articles were highlighted, which means 14.5%, and in 2017, several 24 articles were published, which means 12%. The years 2023 and 2024 have a lower number of highlighted articles, but this is also because the Publish and Perish application also extracts articles based on the number of citations.

Table 2 lists the titles of the journals from which the articles analyzed in the study were extracted, the year in which they were published from 2016 to 2024, and the journal impact factor.

Table 1. Top of the most publications in the period 2016–2024 in the context of sustainability in the pharmaceutical industry.

Year	No. of publication	% of 200
2016	19	9.5%
2017	24	12%
2018	29	14.5%
2019	31	15.5%
2020	36	18%
2021	31	15.5%
2022	15	7.5%
2023	8	4%
2023	7	3.5%
Total	200	100%

Table 2. The list of the journals where the analyzed works were published.

Journal Title	Publication Year	Impact Factor
“Decision Analytics Journal”	2023	3.9
“Journal of Toxicology and Environmental Health”	2017	2.4
“Computers & Industrial Engineering”	2019	6.7
“The Annals of Operations Research”	2021	4.4
“The Journal of Climate Change and Health”	2021	4.8
“African Journal of Science, Technology, Innovation and Development”	2021	1.3
“European Journal of Pharmaceutical Sciences”	2024	4.3
“International Journal of Production Economics”	2016	4.2
“International Journal of Integrated Waste Management, Science and Technology”	2023	7.1
“Sustainable Chemistry and Pharmacy”	2022	5.5
“Journal of Pharmaceutical Policy and Practice”	2023	4.2
“Science of The Total Environment”	2021	8.2

4. Results and discussion

Using the VOSviewer tool, an analysis of the keywords found in the chosen articles from the “Google Scholar” database was performed, based on the number of occurrences of each keyword. An analysis of the keywords used in the selected papers was carried out using the VOSviewer tool, which can be seen in Figure 1.

Following the analysis, 1285 keywords with at least eight occurrences for each keyword were found. In a ranking of the keywords that were most cited, we list six of them: Pharmaceutical industry (129 occurrences), sustainability (71 occurrences), industry (65 occurrences), review (38 occurrences), application (27 occurrences), and development (19 occurrences). In this research area, there are also other keywords with fewer occurrences, such as challenge, research, sustainable development, food, chemicals, waste, analysis, research, food industry, chemical biocatalysis, and chemicals.

These terms generated the highest number of citations between 2019 and 2020 (Figure 2)

Using VOSviewer, a situation of authors who have published works in the same context, of sustainability in the pharmaceutical industry, was also created, which is presented in Figure 3.

According to the literature, the sustainability of pharmaceutical companies can increase when decisions are made that take the environment into account. Doctors, consumers, or drug buyers in hospitals and institutions could make decisions based on the presentation of the environmental effects of active substances. However, incorporating environmental elements into the decision-making process can stimulate and guide organizations to give environmental concerns more attention. The impact that drugs have on the environment can cause patients to worry about whether they

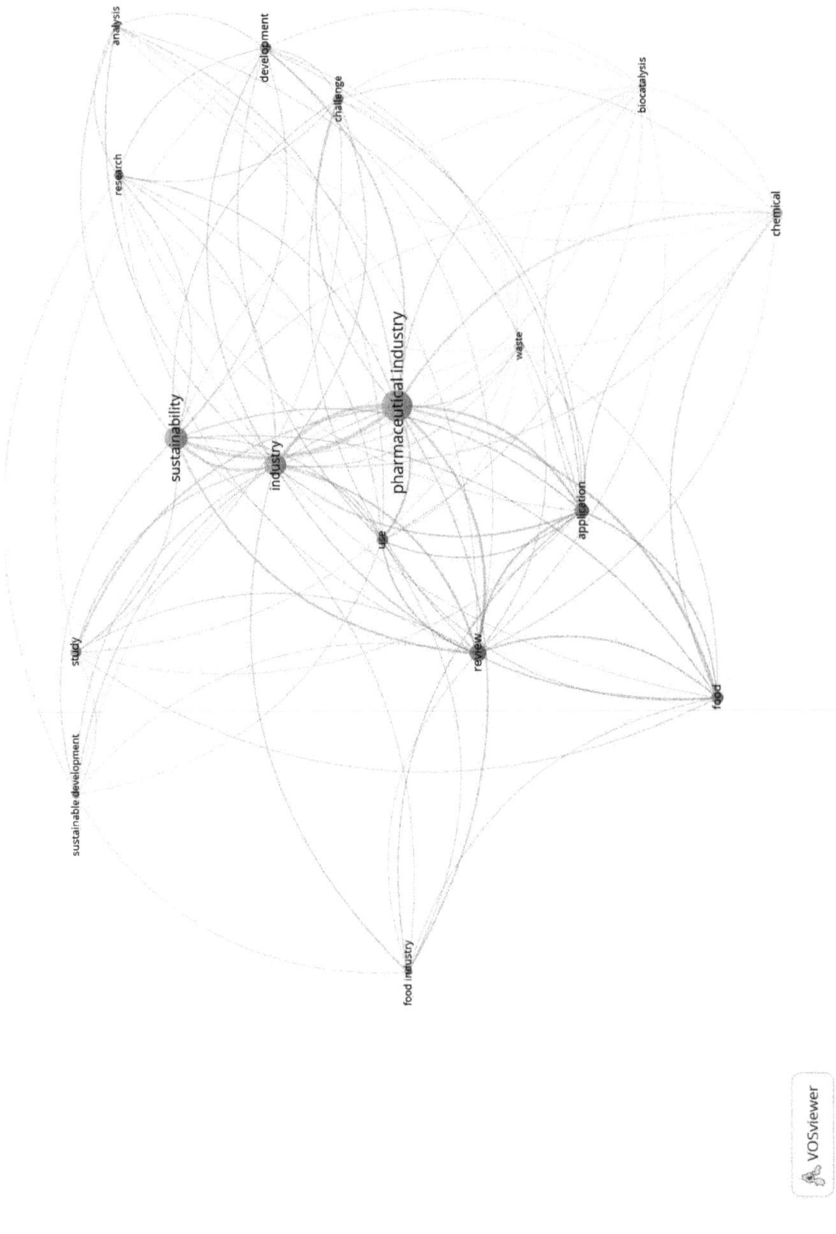

Figure 1. Keyword analysis network.

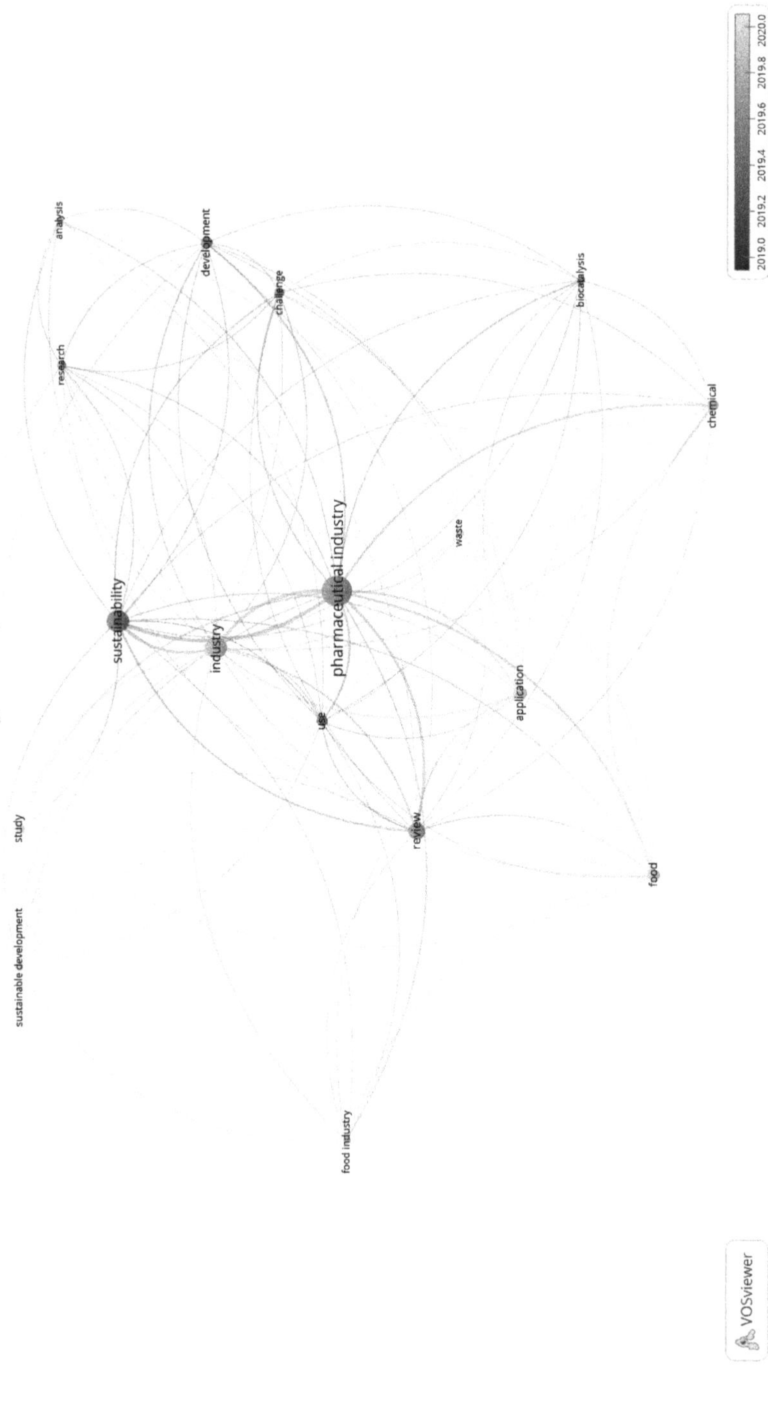

Figure 2. The most frequently used keywords from 2019 to 2020.

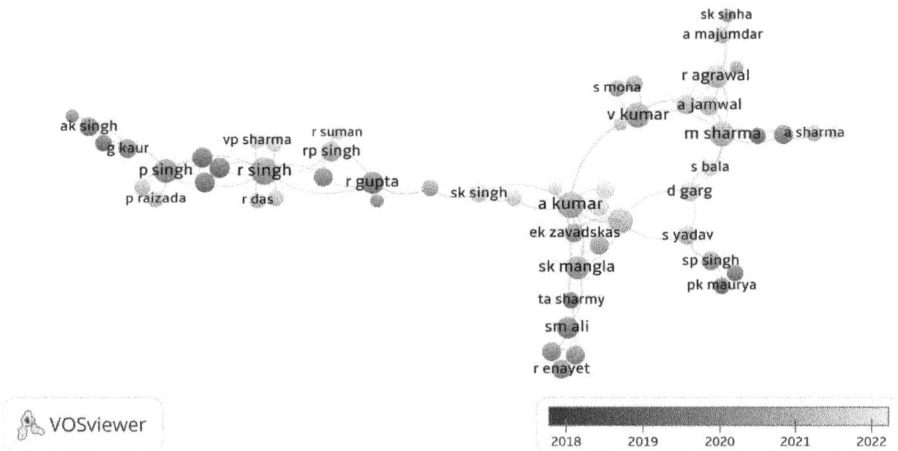

Figure 3. Authors analysis network.

will receive the necessary treatment. On the other hand, perhaps patients should not be involved in choosing their drugs for these reasons. However, recognizing the effects of drugs on the environment and using the most dangerous ones as little as possible could be considered an important step toward increasing sustainability among pharmaceutical companies. To prevent unnecessary environmental pollution, consumers should be better informed about the proper disposal of outdated and unneeded medications. Pharmaceutical companies can support the idea of sustainability through campaigns for public awareness aimed at drug consumers to combat general knowledge of how medications affect the environment. A series of articles were found in the literature that address the challenges that the pharmaceutical industry faces in implementing the concept of sustainability, and Figure 4 presents those that were found in several articles.

The following describes the key challenges in implementing sustainability in pharmaceutical companies, listed above in Figure 4.

1. Legislative challenges include the lack of a legal framework to regulate and require greener operations in the pharmaceutical industry (Wilkinson et al., 2022).
2. Restricting advertising that could prevent customers from learning about prescription drugs can be considered an example of changing national or international legislation (Weraikat et al., 2016).
3. The lack of mandatory and surveillance systems in purchasing and production chains, concentrated production in nations with lax environmental regulations, maintaining the environmental responsibility of suppliers, and the limited opportunities for individual companies or nations to influence global markets and trends are all examples of the lack of information and the need to improve environmental sustainability in production chains (Eriksen et al., 2017).
4. The ecological impact of drug use, the establishment of the most crucial environmental goals for setting priorities, the development of more

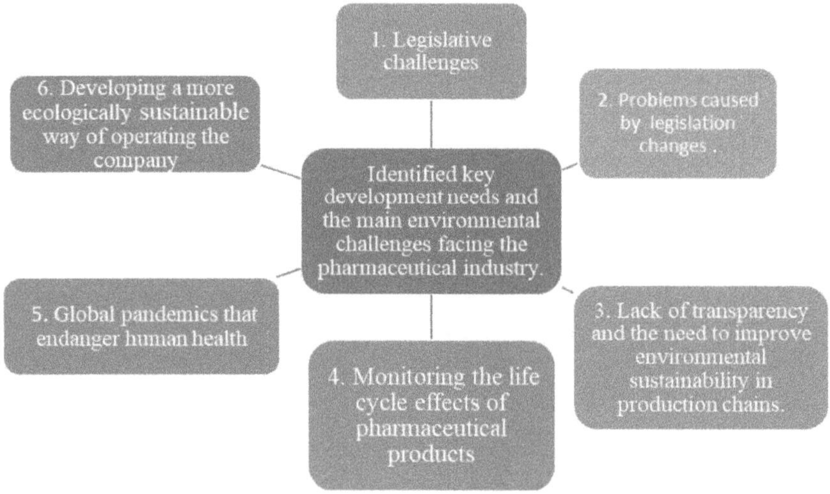

Figure 4. The pharmaceutical industry's primary environmental concerns and critical development needs.

environmentally friendly pharmaceutical products, knowledge of the life cycle impact on environmental sustainability (Sanja Riikonen et al., 2024).

5. In recent years, the variety of types of viruses that lead to the emergence of various pandemics has increased considerably, either due to climate change or antimicrobial resistance (Emmanuel et al., 2022).
6. In pharmaceutical companies, emphasis should be placed on developing an environmental management system that could contribute to changing people's mindsets. Also, environmental assessment of the way of operating and target-oriented development and identification of the requirements and needs of interested parties (Sanja Riikonen et al., 2024).

By identifying the main challenges faced by the pharmaceutical industry in implementing sustainability and basis the information from academic literature, we can then propose solutions to improve or increase the concept of sustainability for pharmaceutical companies. Environmental sustainability is the branch that is considered to be the most difficult to manage due to pharmaceutical products with a high pollution impact on the planet. The active substances that medicines generally contain are quite difficult to eliminate so the residues are harmful to life in general, both humans and other living things, also plants. Raising awareness among people, especially those directly involved in the manufacture of these products, is really important and could have a positive effect on the environment, if the transmission of information about how essential it is to protect what nature gives us is carried out at the level of producers and then workers (people hired to perform the work itself). A very well-informed but also organized environmental management plays an important role in developing sustainable production in pharmaceutical organizations.

The environmental effects of pharmaceuticals should be evaluated in the early preclinical stages of drug design and development to either develop drugs that can be more effectively targeted to their site of action, reducing the dose and thus the amount released into the environment, or prevent the use of toxic substances when possible (Sanja Riikonen et al., 2024). The rational use of medications can be a significant way to reduce the consumption of unnecessary and suboptimal drugs, thereby diminishing the environmental burden and the negative effects of pharmaceutical residues. It has also been observed that doctors should pay attention to the number of medications prescribed at any given time so that patients do not buy too many medications, or at least not more than they need to administer (Sanja Riikonen et al., 2024). On the other hand, those living in disadvantaged areas, such as rural environments, may have limited access to medications or a smaller number of pharmacies to buy extra in the hope of having it on hand in case they need it and thus could be disadvantaged because they cannot afford to visit pharmacies as often as they need to. This leads them to buy extra in the hope of having it on hand in case they need it. In Finland, for example, a legislative decree was amended to require the storage and supply of reserve medicines, which would mean more flexibility, to prevent needless waste from products that have expired.

Another situation of sustainable production could be the inclusion of environmental (production) criteria in the marketing authorization or the "Good Manufacturing Practices (GMP)" rules to guarantee that the effects on the environment are taken into account when producing pharmaceutical products, but in poor countries, regulations, and their execution may be insufficient. Other changes that are considered to be beneficial include legislation that allows organizations to inform about the environmental impact of prescription drugs without interpretation of this as advertising or replacing printed leaflets with electronic ones to lessen the amount of used drug packaging (Sanja Riikonen et al., 2024).

Weraikat et al. (2016) investigate ways to motivate consumers to return unneeded medications. According to the authors, it is not only better for the environment because fewer medications are dispersed, but it also enables the socially responsible distribution of these unneeded medications to parts of people that have limited access to them. Regarding the relationship between economic and environmental sustainability, the studies reviewed provide a model for optimization for the plans that are integrated of manufacturing and supply chain dispersion. Reducing costs and reducing greenhouse gas emissions are the two economic and environmental objectives of the model. In addition, all operations involving the reuse of materials and products in the supply chain are included in the theory of reverse logistics.

Other elements that can support sustainability in the pharmaceutical industry include the rational use of medicines, promoting and supporting the recycling of materials from drug packaging, adopting the use of renewable energy in the company, and solving problems with production emissions.

First, the product life cycle is essential for the manufacturing process. A system based on the product life cycle that considers the sustainability and environmental aspects of the entire product could be considered an effective tool for the decision-making process and to aid in the creation of medications made in a sustainable manner (Kun Wang et al., 2021).

Making production networks more transparent to help consumers make decisions and educating the people about the effects on the environment of pharmaceuticals so that they can consider environmental factors in their decision-making could also be an important basis for promoting the concept of sustainability within pharmaceutical organizations. Raising awareness about appropriate methods of disposing of medicines, as well as the effects on consumers and the general atmosphere in society.

Waste disposal causes not only environmental pollution but also more serious problems, such as the spread of drug-resistant bacteria in the environment. These bacteria are among the greatest threats to public health globally. Waste management is therefore not only an issue that concerns the environment but also economic consequences and social sustainability. Increasingly restrictive legal regulations regarding environmental impacts are evidence of the need for green solutions. Supported by compelling environmental taxes on the rights in question, these regulations address concerns such as trash and the expenses associated with its classification and storage, emissions and discharges, and the challenge of limiting their extent (Chomać-Pierzecka, 2023).

A study in Germany found that after human medical treatment requiring hospitalization, approximately 16,000 tons of pharmaceutical waste are discarded annually. Approximately 60–80% of these pharmaceuticals are disposed of in drainage systems, toilets, landfills, or regular household waste (Melody, 2021). When pharmaceutical waste is improperly disposed of, it might get discharged into the environment , contaminating water, soil, and food. This can affect both wildlife and humans. In addition, the presence of various medicines in the environment can affect the effectiveness of antibiotics, which can lead to the emergence of antibiotic-resistant bacteria.

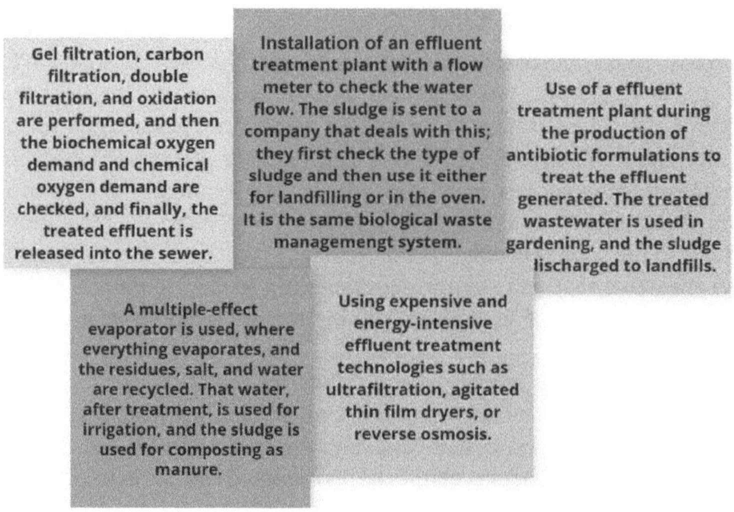

Figure 5. Measures adopted by Indian pharmaceutical companies to support sustainability according to the study conducted by Anita Kotwani et al. (2023).

Although there is little information on the use and waste of medications in many low- and middle-income nations, the dire circumstances in high-income nations may point to a situation that is comparable to that in low-income nations. For example, the consumption of medicines is estimated to increase by 40 to 70 percent in Germany alone by 2045 (Binaya Sapkota and Agamuthu Pariatamby, 2023). Both humans and animals are affected by the consumption of medicines. Veterinary antibiotics used in agriculture, disease treatment, and growth promotion will increase by 67 percent worldwide by 2030 (Sapkota and Pariatamby, 2023). Pharmaceutical use may also rise as a result of overprescribing in clinics and hospitals, consumer self-medication, therapy modifications or changes brought on by incorrect diagnoses, and other medically justified or unjustified causes.

According to a study by Anita Kotwani et al. (2023), in India, several pharmaceutical companies are exposing the measures they have adopted to support sustainability at the company level. Some of the solutions they have adopted are presented in Figure 5.

5. Conclusions

In conclusion, it can be seen that the implementation of sustainability in pharmaceutical companies is a challenge in itself. This industry produces medicines that have beneficial effects on people in the first phase when they treat them, but also negative effects when they are manufactured and then transformed into waste, polluting the environment where humans survive. The use of the "Harzing Publish or Perish" software application made it possible to extract the 200 articles from the "Google Scholar" database, and with the help of the VOSviewer tool, the years with the most publications in the context of sustainability in the pharmaceutical industry were highlighted. Maps were also created with the most common keywords in the selected articles, depending on the number of occurrences of each. From the articles considered for the study, challenges that the pharmaceutical industry faces in implementing sustainability were extracted. In the pharmaceutical industry, particular emphasis is placed on protecting the environment, and one of the important factors in its degradation could be the indiscriminate disposal of pharmaceutical waste. From the literature, possible measures to improve and increase the degree of sustainability in companies in the pharmaceutical industry were extracted. An important measure to support sustainability among pharmaceutical companies could be informing people of the toxicity of pharmaceutical waste and its impact on human health. Doctors play an important role in this process of awareness and information of humanity but also in prescribing medicines for patients, who tend to usually buy more than they need, thus ending up collecting medications that do not have a very long shelf life, which further leads to the creation of waste. The legislation in each country has an important role in encouraging and supporting pharmaceutical companies in implementing this concept of sustainability. Also at a global level, the introduction of strict rules on sustainable production techniques that would influence the authorization of manufacturing and sale could be an important step. Some authors recall this concept of circular economy, which has become very well known in recent years and is applied among pharmaceutical companies. The application of concepts, ideas, and technology to revive consumed or dead resources in order to lessen their negative impacts on the environment, human health, and the economy is known as a circular economy. Timely use of medications during expiration dates is often delayed for various reasons, like pharmaceutical firms' delayed marketing; improper storage, delivery and inventory control procedures used by pharmaceutical firms, medicines distribution channels, government supply networks, as well as community or hospital pharmacies; and even inappropriate use and storage by drug manufacturers. All of these lead to the unintentional expiry of medicines, which is a financial expense that can be largely avoided if there is good waste management so that they can be managed properly.

In the future, the paper could be further developed by conducting a case study that includes concrete data from pharmaceutical companies where this is permissible, as well as sustainability challenges and solutions for implementing sustainability within the organization.

References

Abbasi, Sina and Hassan Ahmadi Choukolaei. (2023). A systematic review of green supply chain network design literature focusing on carbon policy. Decision Analytics Journal 6: 100189.

Ahmadi, Abbas, Mousazadeh, M., Torabi, S. A. and Pishvaee, M. S. (2018). OR applications in pharmaceutical supply chain management. Operations Research Applications in Health Care Management (2018): 461–491.

Anita Kotwani, Ajita Kapur, Mihir Chauhan and Sumanth Gandra. (2023). Treatment and disposal practices of pharmaceutical effluent containing potential antibiotic residues in two states in India and perceptions of various stakeholders on contribution of pharmaceutical effluent to antimicrobial resistance: a qualitative study. Journal of Pharmaceutical Policy and Practice 16: art.59.

Bhatta, D. D., Sarfraz, M., Ivascu, L. and Pislaru, M. (2023). The nexus of corporate affinity for technology and firm sustainable performance in the era of digitalization: a mediated model. Sustainability 15(12): 9765. https://doi.org/10.3390/su15129765.

Binaya Sapkota and Agamuthu Pariatamby. (2023). Pharmaceutical waste management system – Are the current techniques sustainable, eco-friendly and circular? A review. International Journal of Integrated Waste Management, Science and Technology 168: 83–97, https://1610asrbp-y-https-doi-org.z.e-nformation.ro/10.1016/j.wasman.2023.05.052.

Brodin, Tomas, Johanna Nordling, Annelie Lagesson, Jonatan Klaminder, Gustav Hellstrom, Bent Christensen et al. (2017). Environmental relevant levels of a benzodiazepine (oxazepam) alters important behavioral traits in a common planktivorous fish. Journal of Toxicology and Environmental Health, Part A 80(16-18): 963–970.

Brundtland, Gro Harlem. (1987). Our common future—Call for action. Environmental Conservation 14.4: 291–294.

Chomać-Pierzecka, E. (2023). Pharmaceutical companies in the light of the idea of sustainable development—an analysis of selected aspects of sustainable management. Sustainability 15(11): 8889. https://doi.org/10.3390/su15118889.

Darestani, Soroush Avakh and Mehran Hemmati. (2019). Robust optimization of a bi-objective closed-loop supply chain network for perishable goods considering queue system. Computers & Industrial Engineering 136: 277–292.

Elkington, John and Ian H. Rowlands. (1999). Cannibals with forks: The triple bottom line of 21st century business. Alternatives Journal 25.4: 42.

Emmanuel Sunday Okeke, Timothy Prince Chidike Ezeorba, Cherles Obinwame Okoye, Yoo Chen, Guanghua Mao et al. (2022). Environmental and health impact of unrecovered API from pharmaceutical manufacturing wastes: A review of contemporary treatment, recycling and management strategies. Sustainable Chemistry and Pharmacy 30, https://1610asrbp-y-https-doi-org.z.e-nformation.ro/10.1016/j.scp.2022.100865.

Eriksen, J., Gustafsson, L. L., Ateva, K., Bastholm-Rahmner, P., Ovesjö, M. L., Jirlow, M. et al. (2017). High adherence to the 'Wise List' treatment recommendations in Stockholm: a 15-year retrospective review of a multifaceted approach promoting rational use of medicines. BMJ Open 7(4): e014345.

Farhad Safaei and Naeme Zarrinpoor. (2024). Design of a pharmaceutical supply chain in uncertain conditions considering financial strategies and environmental concerns. Operation Management Research 17: 891–915. https://doi.org/10.1007/s12063-024-00440-0.

Fernandes, A. G. (2016). Closing the loop-the benefits of the circular economy for developing countries and emerging economies. Teddington: Tearfund.

Goodarzian, F., Ghasemi, P., Gunasekaren, A., Taleizadeh, A. A. and Abraham, A. (2021). A sustainable-resilience healthcare network for handling the COVID-19 pandemic. The Annals of Operations Research 1–65.

Gower, R., Schröder, P., Khinmaung-Moore, J., Cook, P., Ling, A., Green, J. et al. (2016). Virtuous circle: How the circular economy can create jobs and save lives in low and middle-income countries.

Hulea, M., Rosu, O., Miron, R. and Aştilean, A. (2018). Pharmaceutical cold chain management: Platform based on a distributed ledger. International Conference on Automation, Quality and Testing, Robotics (AQTR) (pp. 1–6). IEEE.

Kun Wang, Tao Zhuang, Zhaoxin Su, Menghao Chi and Haichao Wang. (2021). Antibiotic residues in wastewaters from sewage treatment plants and pharmaceutical industries: Occurrence, removal and environmental impacts. Science of The Total Environment 788, https://1610asrs4-y-https-doi-org.z.e-nformation.ro/10.1016/j.scitotenv.2021.147811.

Larsson, D. G. Joakim. (2014). Pollution from drug manufacturing: review and perspectives. Philosophical Transactions of the Royal Society B: Biological Sciences 369.1656. https://doi.org/10.1098/rstb.2013.0571.

Marius, P., Elena-Lidia, L. D. A., Ionut-Viorel, L. D. H. and Petronela, L. D. L. R. (2017). Fuzzy logic-a practical tool for monitoring urban areas environmental performance. International Multidisciplinary Scientific GeoConference: SGEM; Sofia 17. DOI:10.5593/sgem2017/52.

Melody Okereke. (2021). How pharmaceutical industries can address the growing problem of climate change 2021. The Journal of Climate Change and Health 4: 100049. https://1610a8e0c-y-https-doi-org.z.e-nformation.ro/10.1016/j.joclim.2021.100049.

Ofori, Peres. (2021). Waste disposal sites and residential rental values nexus: an appraisal of agogo asante akyem dumps. African Journal of Science, Technology, Innovation and Development 13.2: 223–233.

Ofori, P. (2023). Circular economy practices in third world nations: challenges and implications for environmental sustainability. Environ Dev Sustain. https://doi.org/10.1007/s10668-023-03939-x.

Pislaru, M., Herghiligiu, I. V. and Robu, I. B. (2019). Corporate sustainable performance assessment based on fuzzy logic. Journal of Cleaner Production 223: 998–1013. https://doi.org/10.1016/j.jclepro.2019.03.130.

Sanja Riikonen, Johanna Timonen and Tiina Sikanen. (2024). Environmental considerations along the life cycle of pharmaceuticals: Interview study on views regarding environmental challenges, concerns, strategies, and prospects within the pharmaceutical industry. European Journal of Pharmaceutical Sciences 196, https://1610asaet-y-https-doi-org.z.e-nformation.ro/10.1016/j.ejps.2024.106743.

Silva, A. C., Marques, C. M. and de Sousa, J. P. (2023). A simulation approach for the design of more sustainable and resilient supply chains in the pharmaceutical industry. Sustainability 15(9): 7254. https://doi.org/10.3390/su15097254.

Syed Ale Raza Shah, Qianxiao Zhang, Jaffar Abbas, Daniel Balsalobre-Lorente and Ladislav Pilar. (2023). Technology, urbanization and natural gas supply matter for carbon neutrality: A new evidence of environmental sustainability under the prism of COP26. Resources Policy 82: 103465.

Wan Te Liew, Arief Adhitya and Rajagopalan Srinivasan. (2014). Sustainability trends in the process industries: A text mining-based analysis. https://1610a8l15-y-https-doi-org.z.e-nformation.ro/10.1016/j.compind.2014.01.004.

Weraikat, D., Zanjani, M. K. and Lehoux, N. (2016). Two-echelon pharmaceutical reverse supply chain coordination with customers incentives. International Journal of Production Economics 176: 41–52. https://doi.org/10.1016/j.ijpe.2016.03.003.

Wilkinson John L., Alistair B. A. Boxall, Dana W. Kolpin, Kenneth M. Y. Leung , Racliffe W. S., Christobal Galban-Malagon et al. (2022). Pharmaceutical pollution of the world's rivers. Proceedings of the National Academy of Sciences 119.8: e2113947119.

Chapter 12

An Innovative Concept for an Eco-Friendly Sustainable Approach within Organisations

Digital Green Lean Six Sigma

Nicoleta-Mihaela Căsăneanu (DASCĂLU), * *Adrian Vilcu,*
Laura Crina Miraute (COCA) and *Ema Alexa*

1. Introduction

Major shifts in global average temperature, precipitation patterns, the extinction of once-common species, and human health have all resulted from the increase in greenhouse gases. From 1990 to 2023, there was a 45% surge in greenhouse gas emissions. Almost 20 percent of all greenhouse gas emissions come from industrial organizations, which use natural resources at an uncontrolled rate and don't have adequate waste management systems in place. To further benefit the company and its workers, automation of the manufacturing process is being pursued. This will make the process easier and allow for a multiplication of the quantity of the finished product (Coca and Pîslaru, 2022). Sustainable practices such as Green Lean Six Sigma (GLSS) can help industries combat this. This method is advantageous, positively impacting the long-term environmental and financial success of a corporation. To address climate change and create a more equitable and sustainable world, we must re-evaluate the international economic system (Ma et al., 2021; Cherrafi et al., 2022). Green efforts have garnered considerable attention from

"Gheorghe Asachi" Technical University of Iasi-Romania, Faculty of Industrial Design and Business Management, Iasi, Romania.

* Corresponding author: nicoleta-mihaela.dascalu@student.tuiasi.ro

DOI: 10.1201/9781003518990-12

scholars and policymakers, who are emphasizing greenhouse gas emissions. Advancements in sustainable technologies and practices can alleviate the adverse effects of economic development on the environment. Businesses must dedicate themselves to endorsing green projects to enhance their financial and environmental performance. Businesses undertake green investment, an economic activity and intricate management process, to attain a "win-win" scenario that enhances both financial outcomes and ecological impact (Antonioli et al., 2022; Edwin Cheng et al., 2022; Neligan et al., 2023). Green Lean Six Sigma (GLSS) is a sustainable approach to enhance process efficiency by reducing waste, unpredictability, and pollution (Dues et al., 2013; Garza-Reyes, 2015). By integrating with green technology, GLSS can improve material efficiency and reduce negative environmental impacts. An individual organization can incorporate GLSS, fulfil consumer demands, and promote sustainability. The initial step involves identifying a project with the greatest potential for long-term sustainability through techniques such as environmental value stream mapping and life cycle evaluation (Garza-Reyes et al., 2018). The third step involves examining the project's inefficiencies and waste through techniques such as five-why analyses and cause-and-effect diagrams. The fourth phase encompasses documenting activities, teaching personnel on the optimal identified solution, and proposing, assessing, and executing the best solution (Ruben et al., 2018). New technologies, integrating Lean and Six Sigma methodologies, possess the capacity to transform manufacturing and various other sectors (Kamble et al., 2020; Rosa et al., 2020; Skalli et al., 2023). It encompasses machine learning, the Industrial Internet of Things (IIoT), artificial intelligence (AI), big data analytics, and various other digital technologies. Businesses can enhance cost performance, safety, quality, delivery, and efficiency through the implementation of these technologies (Di Maria et al., 2022). The implementation of Lean and Six Sigma principles in the contemporary digital landscape requires a team of proficient leaders, managers, and employees. The IIoT collects real-time data on machine uptime, performance, and downtime, among other essential metrics, allowing organizations to enhance OEE, pinpoint areas for kaizen campaigns, and more (Rosin et al., 2020). Six Sigma methodologies can utilise AI and machine learning to enhance comprehension of the critical process parameters related to variation and process yields (Ciano et al., 2020). By analyzing extensive datasets, firms can attain a more profound understanding of their operations and focus on critical performance and quality indicators. In addition to identifying essential process drivers, artificial intelligence and machine learning may help firms optimize production schedules to reduce waste and enhance productivity (Romero and Flores, 2019). Organizations can enhance their profitability by employing targeted enhancements through these technologies and Six Sigma methodologies. Analytics of big data can enhance Lean and Six Sigma methodologies by identifying the sources of process variation along the entire value chain (Saabye et al., 2020). To attain continuous improvement, firms can investigate the causes of process deviations, identify opportunities for enhancement, and collect and analyse extensive data sets (Tortorella et al., 2019). Businesses can enhance overall equipment effectiveness (OEE) by utilizing big data to anticipate equipment failures and schedule maintenance accordingly. As a supplementary element of their commitment to the leadership standard for tiered reviews, it can aid in developing and tracking key performance indicators (KPIs) essential to their success. Combining digital twin technology with Lean and Six Sigma methods enhances output, increases consistency, and boosts efficiency by mimicking the functioning of real assets in a digital environment (Shamsi and Alam, 2018). Enhancing quality and minimizing errors is possible through data-driven decisions facilitated by digital twins, which can simulate various scenarios, identify production bottlenecks, optimize material and product flow, and more. Lean and Six Sigma environments can use digital twins for predictive maintenance, forecasting equipment failures, and proactively arranging maintenance schedules. The efficacy of new technology initiatives inside a Lean and Six Sigma framework depends on the integrator role. Integrating traditional methods with advanced technologies necessitates a specific skill set, knowledge, and personal experience. For organizations to prosper in this field, they must invest in elite talent and foster a growth mindset. This will enable them to optimize the advantages of new technologies in terms of efficiency, effectiveness, quality, and cost (Alexa et al., 2023). An increasing number of enterprises are adopting sustainable practices to minimize their ecological impacts and enhance environmental efficiency. This technique encompasses the 3 Rs of production—reduce, reuse, and recycle—alongside sustainable supply chain management. By employing this strategy, organizations can mitigate their environmental impact and ecological footprint without jeopardizing their financial objectives (Mathiyazhagan et al., 2019; Khan et al., 2023; Abbasi and Choukolaei, 2023). GLSS has emerged in the digital innovation era as a corporate approach that incorporates the 3 Rs—reduce, reuse, and recycle—to support the circular economy. Notwithstanding its acknowledged advantages in the industrial sector, practitioners continue to exercise caution regarding its implementation due to insufficient awareness and cultural factors. It is essential to identify the crucial success factors for this green initiative and to structure the existing body of knowledge accordingly (Letchumanan et al., 2022). However, the existing literature lacks sufficient information on how to integrate the latest industry technology with GLSS principles to expedite the sustainable transformation of the eco-friendly manufacturing sector. By investigating how digital innovations can successfully improve GLSS and contribute to more environmentally friendly manufacturing practices, this research aims to fill this knowledge gap. This research investigation aims to identify and assess potential relationships between emerging technologies and existing GLSS strategies, with the objective of assisting factories in their efforts to become environmentally friendly. This approach analyses the latest research and enhances it with additional conceptual frameworks, demonstrating how modern technologies can enhance GLSS. Through their findings, professionals will gain a roadmap to implement GLSS strategies that leverage modern technologies, thereby making their organisations more environmentally friendly. Consequently, our study addresses two main research questions:

RQ1: Can emerging technologies play a pivotal role in eco-friendly manufacturing through the adoption of GLSS strategies?

RQ2: Does the GLSS and digital innovation have a direct impact on eco-friendly organisations?

To meet the study objectives and tackle the research issues mentioned above, the conceptual research model illustrated in Figure 1 was used. The current study aims to examine the connection between modern technological advances and the implementation of green and lean Six Sigma methodologies. This research highlights the degree to which innovative technologies promote three particular approaches and offers perspectives on how manufacturers can effectively leverage these technologies to enhance environmental responsibility. The presented research introduces a new, thorough method for analysing the interactions between green,

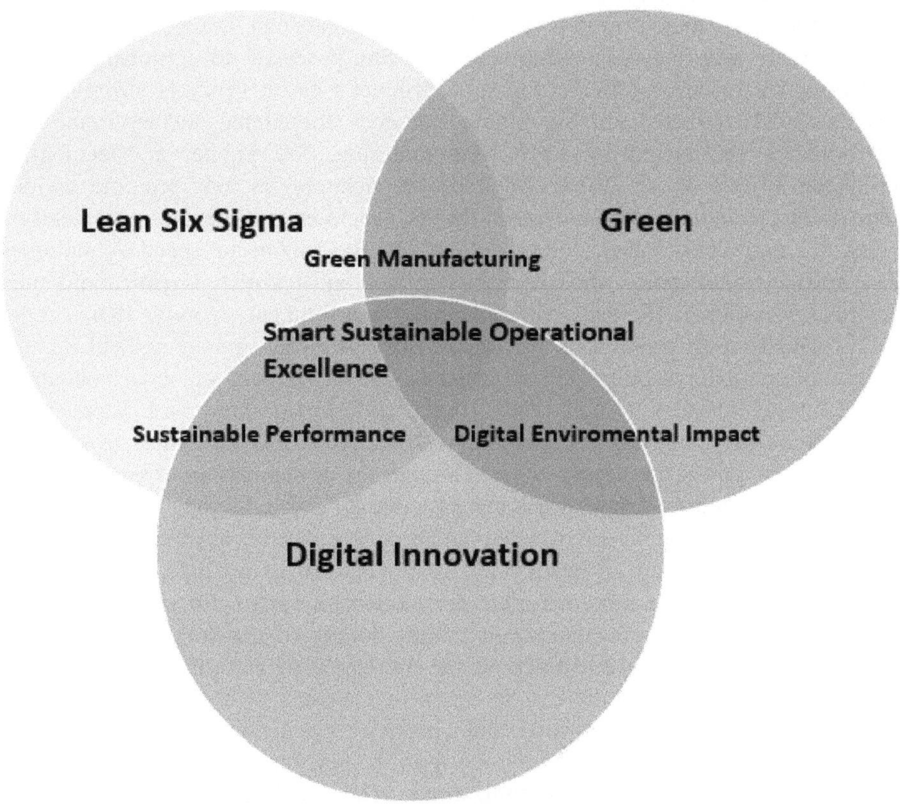

Figure 1. Convergence between lean six sigma, green and digital innovation.

LSS, and digital innovations within the framework of environmentally conscious manufacturing performances. The analysis delves into the interconnected impact of these three concepts, offering crucial insights into how their collective application can improve environmental friendly industrial performance, thereby addressing a significant research need. The paper makes an original contribution by providing concrete recommendations for professionals in the industry on how to successfully incorporate and use green principles, LSS tools, and cutting-edge technologies to enhance environmental friendly production performances. The structure of the paper's subsequent sections is as follows: Section 2 outlines the theoretical concepts, hypotheses, rises, results, and discussions. Section 3 provides a framework for businesses to align their operational strategies with environmental objectives and a detailed examination of the results. Section 4 provides implications and theoretical contributions for the study, along with a discussion of limitations and the agenda for future research. Section 5 presents a final conclusion.

1.1 Integrating green lean six sigma with digital innovation

Contemporary firms are increasingly favoring the acquisition of sustainable goods and services, as they prioritize social and environmental issues. Industries are progressively adopting sustainable manufacturing processes to diminish pollution and optimize resource efficiency over a product's entire lifecycle (Miraute and Pîslaru, 2024). Green Lean Six Sigma enhances operational and environmental performance by integrating Lean Manufacturing, Six Sigma, and ecological principles (Yacob et al., 2019). Advances in technologies may promote various objectives, including human-centric design, socio-environmental sustainability, resilience, and sophisticated ecological solutions. By utilizing advanced technologies like artificial intelligence, the internet of things, and robotics, organizations may improve operational efficiency while reducing waste and energy usage (Khan et al., 2023). This synergy promotes a sustainable manufacturing strategy and cultivates a culture of continual development, resulting in a more resilient and environmentally conscientious industry (Gatell and Avella, 2023). As enterprises adopt these innovations, they may also collaborate with suppliers and customers to enhance sustainability across the supply chain. Through the dissemination of best practices and creative solutions, corporations can generate a ripple effect that enhances their beneficial influence on the environment and propels collective advancement towards a more sustainable future (Belhadi et al., 2020; Ghaithan et al., 2021; Kamble et al., 2020). This cooperative approach enhances relationships across the supply chain and cultivates a feeling of collective accountability among all stakeholders. When each member adheres to sustainable practices, the entire industry can collectively diminish its carbon footprint and improve its resilience against changing environmental problems. By adopting innovation and openness, businesses may motivate their partners and customers to embrace eco-friendly practices, fostering a culture of sustainability that transcends organizations. This dedication to environmental stewardship benefits the planet and establishes companies as leaders in a swiftly evolving economy. Analyzing how the use of advanced technologies enhances collaboration between workers and machines, results in increased efficiency and decreased waste. This collaborative method not only increases productivity but also promotes a more sustainable environment by emphasizing on the responsible resource usage and reducing the ecological footprint of manufacturing operations (Skalli et al., 2023). This collaboration between human skill and machine accuracy facilitates real-time modifications in manufacturing and enhances resource distribution and energy usage. Consequently, firms can attain increased productivity while upholding their dedication to sustainability, ultimately benefiting both the economy and the environment. By adopting this creative strategy, firms improve their operational efficiency and promote a more sustainable future. This dedication to sustainability can enhance company reputation, attract environmentally conscientious consumers, and establish a competitive edge in a progressively eco-aware market. Moreover, organizations that emphasize sustainable practices frequently discover new avenues for collaboration and innovation, resulting in the creation of advanced technologies that further mitigate their environmental impacts. This cultivates a culture of accountability and stewardship, motivating employees and stakeholders to actively participate in sustainability activities (Anass et al., 2021; Buer et al., 2018; Kamble et al., 2020). Cutting-edge technology and methodologies like GLSS can accomplish sustainable development and resource preservation. In contrast to Six Sigma's focus on reducing process variability and defects, Lean aims to maximize process flow while minimizing waste. Green operations encompass the entire product lifecycle, comprising eco-design, sustainable manufacturing, recycling, reuse, distribution, use, and disposal of products. Businesses can achieve important goals like reducing waste, mitigating pollution, optimizing processes, conserving energy and resources, and improving cost efficiency by combining green Lean Six Sigma with cutting-edge systems (Mittal et al., 2026; Kaswan et al., 2023). Spending reduction is one of these goals. Modern technology and GLSS are converging, which means there are chances to combine the two for improved operational efficiency and less environmental impact (Boumsisse and Benhadou, 2024). While cutting-edge technology seeks to improve material flows and energy efficiency, GLSS places an emphasis on optimizing processes to lessen environmental impacts and eliminate waste. To achieve sustained success, enterprises may use digital transformation strategies that include AI, IoE, cloud computing, big data, cyber-physical systems, 5G, and additional technologies. Advancements in energy management and carbon emission mitigation are integral to advances in technologies. These encompass the Intelligent Energy Management System (IEMS), microgrids, cloud-based demand response systems, advanced storage and charging solutions, smart grids, and blockchain-enabled peer-to-peer energy trading (Pandey et al., 2018). New technological developments focus on affordable mass customization with pinpoint accuracy, emphasizing an adaptable mindset against the backdrop of human-focused technological advancement. This aligns with GLSS goals of reducing overproduction and excess inventories, making it an advantageous decision. Digital twins are a game-changer for product customization because they let businesses model and forecast the digital socio-environmental impacts of their goods and services across their whole lifespan, from design to disposal. Green operations encompass the entire product lifecycle, comprising eco-design, sustainable manufacturing, recycling, reuse, distribution, use, and disposal of products. By integrating GLSS with advanced technologies, companies can attain critical objectives, such as waste reduction, pollution mitigation, process optimization, energy and resource conservation, and cost efficiency. These objectives encompass reducing expenditures. The convergence of GLSS and advanced technologies presents opportunities to integrate operational efficiency and sustainability with modern technologies, as illustrated in Table 1. GLSS prioritizes process optimization to reduce environmental impacts and eradicate waste, whereas advanced technologies aim to enhance material flows and energy efficiency (Kumar et al., 2016). To achieve sustained success, enterprises may use digital transformation strategies that include AI, IoE, cloud computing, big data, cyber-physical systems, 5G, and additional technologies. Advancements in energy management and carbon emission mitigation are integral to advances in technologies (Suresh et al., 2024). These encompass the Intelligent Energy Management System (IEMS), microgrids, cloud-based demand response systems, advanced storage and charging solutions, smart grids, and blockchain-enabled peer-to-peer energy trading. Advances in technologies focus on high-precision, low-cost mass customization, emphasizing resilient thinking within the context of human-centered digital transformation and bioeconomic principles. This aligns with GLSS goals of reducing overproduction and excess inventories, making it an advantageous decision. The implementation of digital twins in product customization is essential, as it enables organizations to simulate and predict the digital socio-environmental effects of their products and services over their entire lifecycle, from creation to disposal. The Internet of Everything (IoE) is improving

Table 1. Interlink between cutting-edge technologies and GLSS.

Environmental metrics	Grean Lean Six Sigma	Advanced technologies
Emissions reduction and energy efficiency	Trimming the climate footprint of industrial workflows Improving consumption energy	Management System (IEMS) smart storage renewable microgrids self-repair
Sustainable process	Reduce waste Minimization environmental impact Process optimization	Artificial intelligence Internet of Things Smart Additive Manufacturing
Sustainable Personalization	Reduction excess inventory Eliminate waste from overproduction	Digital twins Customization of products
Data analysis for sustainable development	Data interpretation and transformation Optimisation Decision-making Organization monitoring environmental and operational performance	Big Data Internet of Things Artificial intelligence
Sustainable Innovation	Sustainable eco-designed products Sustainable logistics Efficient production energy	Digital manufacturing transformation

efficiency and data-informed decision-making, while big data analytics facilitate real-time decision-making, both of which are fundamental components of advanced technologies. Establishing a trustworthy method for gathering relevant data is key to the GLSS approach. Fundamental to cutting-edge tech, sustainable innovation seeks to transform the sector into one that is creative, robust, socially conscious, and competitive—all while respecting and reducing its influence on the environment and meeting international standards. Efforts to create better, more environmental friendly products have garnered substantial investment from the manufacturing sector's research and development program (Cherrafi et al., 2017).

Following are the four primary domains in which these technologies have had a significant positive impact, as identified by the research:

1. Their efforts have resulted in a significant reduction in their carbon footprint. They are optimizing their processes using resources such as artificial intelligence and automation.
2. The use of environmentally friendly materials in manufacturing processes is crucial to shifting toward responsible resource consumption practices. Technological advancements encourage this transition by making it easier to select materials with low environmental impact.
3. Companies maximize waste management, proving how technology helps lower waste generation. This is consistent with the ideas of GLSS, which give resource economy top priority.
4. Adoption of energy-efficient technologies. This highlights the effort to reduce energy usage and encourage the adoption of clean energy sources. This is in line with the GLSS commitment to energy efficiency and a reduction in greenhouse gas emissions.

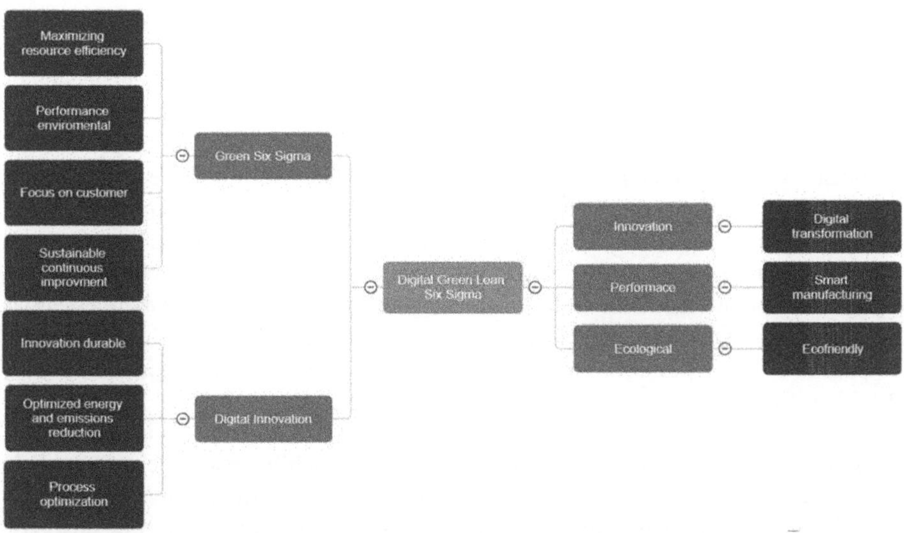

Figure 2. Proposed supply chain frameworks that directly impact eco-friendly organizations.

Emerging technologies not only foster innovation but also examine their impact on business innovation. Furthermore, new technologies have significantly contributed to the sustainability of their operations.

Figure 2 shows how GLSS and cutting-edge technology work together to create sustainable supply chain frameworks, better production methods, and revolutionary, environmentally friendly products.

2. Results and discussion

In the intricate and evolving landscape of the economy, it is imperative for organizations to find equilibrium among productivity, profitability, and sustainability. This balance is essential for generating enduring value for all stakeholders and securing our quest for ongoing success in a world marked by constant change. Public consciousness and apprehension about the impact of industrial activities on the environment and climate change have significantly elevated the notion of sustainability within the manufacturing sector. Consumers are increasingly insisting on products crafted using sustainable methods and materials, and organizations that do not align with these expectations jeopardize their market position. For sustained profitability, it is essential that organizations adopt ongoing enhancement strategies, allocate resources toward emerging technologies, cultivate groundbreaking products and services, and investigate untapped markets. Establishing a robust brand and reputation capable of enduring economic fluctuations and shifts in consumer preferences is of paramount importance. Methodologies such as LSS, Circular Economy (CE), and advances in technologies significantly enhance sustainability efforts, influencing the production chain (Rosa et al., 2020). Innovative technologies can enhance the efficacy of Lean Six Sigma (LSS) and Continuous Improvement (CE) in manufacturing by delivering real-time data and analysis that facilitate the identification and elimination of waste and defects in production (Bai et al., 2020; Lu et al., 2022). For instance, sensors and various Internet of Things (IoT) devices facilitate the collection of data regarding equipment performance, product quality, material usage, energy consumption, and waste generation. We can subsequently analyze this data to identify enhancement opportunities. Cutting-edge technologies like artificial intelligence and machine learning possess the capability to automate and enhance various processes, thereby diminishing the likelihood of human error and elevating overall efficiency. The goal of predictive maintenance algorithms is to find and fix potential equipment problems before they happen. This cuts down on downtime, the need for replacement parts, and overall productivity. The concepts of GLSS, CE, and advanced technologies are significant operational methodologies and organizational practices that are essential for achieving sustainable organizational performance (Ciliberto et al., 2021). LSS emphasizes the importance of minimizing waste and enhancing quality through the identification and eradication of defects (Skalli et al., 2022a). Through the implementation of LSS, organizations can detect and eradicate inefficiencies within their production processes, leading to enhanced sustainability and efficiency in their operations. Advanced technologies denote a collection of digital instruments and technologies that influence all value chains associated with data gathering, analysis, storage, and communication. Introducing advanced technologies to an operation significantly enhances its environmental and operational sustainability, particularly by reducing costs, boosting productivity, optimizing resource utilization, enhancing customer satisfaction, ensuring on-time delivery, and producing high-quality products. Advanced technologies have the potential to address various environmental challenges, thereby fostering sustainable results. Its main goal is to get rid of waste and make processes as consistent as possible. Manufacturing firms have been acquiring insights into methodologies such as lean manufacturing, Six Sigma, and environmentally sustainable practices through an integrated perspective (Cherrafi et al., 2017). Nevertheless, the realms of circular economy and innovative technologies remain nascent, with their integration into the manufacturing sector still at a preliminary stage. In recent years, these concepts have gained significant attention as promising avenues for transforming organizational processes and enhancing resource efficiency. The integration of GLSS and CE presents a compelling opportunity for the advancement of sustainable and efficient business practices (Ciliberto et al., 2021). This amalgamation has the potential to assist organizations in refining their processes while simultaneously mitigating their ecological footprints. Through the adoption of LSS strategies and circular economy practices, organizations have the potential to enhance their processes, minimize waste, and play a pivotal role in fostering a more sustainable future. Improving the environmental performance of companies depends on the investments in new technologies; thus, their relevance is growing. These expenditures demonstrate the ability of businesses to enhance the effectiveness of GLSS. Using predictive analytics, intelligent automation, and real-time monitoring, among other tools, marks a change toward more environmentally friendly and efficient operations. Thanks to new technologies, which have proved to be a strategic lever for sustainable innovation, companies are now able to satisfy growing consumer expectations for environmental friendly products and services as well as regulatory requirements. Conversely, the apparent influence of these technologies raises questions about the potential and challenges associated with their acceptance. Lack of technical knowledge, opposition to change, significant initial implementation expenses, and high implementation costs are among the most significant challenges. The company should use a personalized support plan to go beyond these challenges and guarantee the effective integration of modern technologies. Investing in ongoing staff training is crucial if one is to fully maximise the possibilities of contemporary technology and follow GLSS values. An indispensable requirement for companies is developing a culture of constant improvement and knowledge that supports and enables sustainable innovation and development. The effective application of GLSS ideas and new technologies usually depends on close cooperation among several departments inside a company as well as among the suppliers, consumers, and industrial partners. The spread of knowledge and suggestions for best practices could accelerate the application of these technologies and intensify their beneficial effects on durability.

3. Sustainability through DIGITAL GREEN Lean Six Sigma: A road map to more eco-friendly company operations conclusions

The synergy that exists between these technologies and the practices of GLSS presents an opportunity to enhance operational excellence while simultaneously strengthening the commitment to sustainability. It also offers a framework (Figure 3) for businesses seeking to align their operational strategies with environmental objectives.

Sustainability in manufacturing has emerged as an important goal for industries across the globe in response to growing concerns about the environment. More and more businesses are switching to green production methods in an effort to lessen

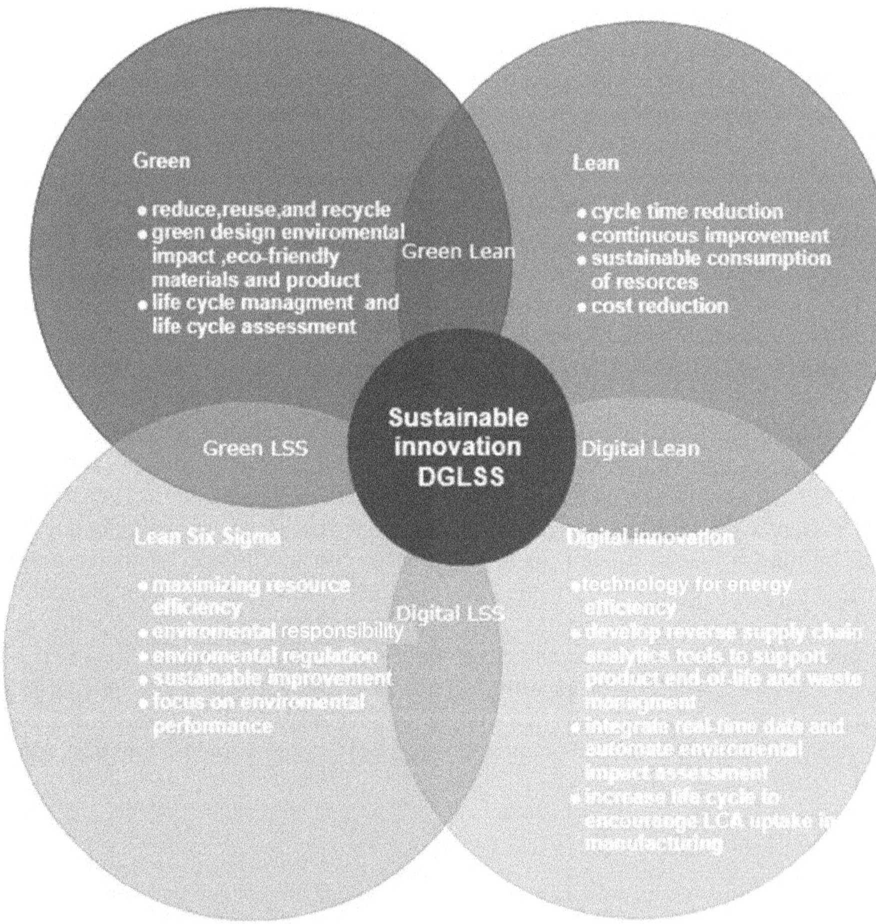

Figure 3. The synergy between new technologies and the practices of Green Lean Six Sigma.

their negative effects on the environment, including energy consumption and carbon emissions. Sustainable manufacturing is more than just a competitive advantage in today's market; it is an integral component of corporate responsibility.

Therefore, we propose the following strategies for positively eco-friendly organizations:

➢ The Value of Eco-Friendly Production Practices

There is increasing pressure on manufacturers to implement sustainable practices that lessen their impact on the environment by cutting emissions, conserving resources, and minimizing damage. These days, eco-friendly production methods are essential to running any successful business. Businesses are realizing that sustainable practices are essential to their long-term profitability as consumer preferences shift towards products with a smaller environmental impact. These initiatives contribute to enhancing the brand›s image, guaranteeing adherence to environmental rules, and drawing in an increasing number of environmentally concerned buyers.

Solar and wind power, which are renewable energy sources, are attracting a growing amount of investment from manufacturers. Businesses can improve their energy independence and drastically cut their carbon footprints by switching to these environmentally friendly options. Energy-efficient machinery and green building designs greatly assist in reducing energy use. In addition, manufacturers are able to quickly identify inefficiencies and make adjustments thanks to smart factory systems that monitor and manage energy consumption in real time.

➢ Recycling and Reducing Waste

Sustainable manufacturing relies on efficient waste management. Producers can lessen their impact on the environment by adopting a circular economy model that emphasizes material reutilization, recycling, and the discovery of novel applications for waste products. Reduced environmental impact, savings on raw materials, and less waste going to landfills are all benefits of this model. Companies committed to zero-waste manufacturing are innovating by recycling and repurposing materials to reduce waste and save costs.

➢ Eco-Friendly Distribution Networks

In addition to a sustainable production process, a sustainable supply chain is critical. Businesses are teaming up with vendors that share the values of using sustainable resources and fair labor practices. This group's combined efforts lessen a product's environmental impact at every stage. To better manage logistics and source materials, many manufacturers are turning to digital platforms that measure the environmental effect of their supply chains.

➢ Technologies for Environmentally Friendly Production

Advanced manufacturing technologies are reducing waste and energy consumption. Manufacturers are now able to use materials more precisely thanks to innovations like additive manufacturing and 3D printing, which significantly reduce excess waste in production. The use of robotics and automation is also crucial for improving production processes, which in turn reduces energy consumption and maximizes efficiency.

➢ Energy Efficiency Initiatives

Manufacturers are putting money into carbon capture technologies to take CO_2 out of their processes before it reaches the air, further reducing their carbon emissions. A growing number of businesses are also investing in renewable energy and reforestation initiatives as part of carbon offset programs, which aim to balance out their emissions.

➢ Managing the Lifecycle of a Product

To lessen the effect that a product has on the environment, sustainable product design is essential. Manufacturers can lessen product waste by designing items with greater ease of disassembly, recycling, or repurposing. Another way to lessen the impact on the environment is to use sustainable materials. These might include biodegradable plastics or recycled metals. To improve sustainability all the way through a product's lifecycle, manufacturers can use lifecycle assessments (LCAs) to measure the effects of their products on the environment at each step, from design to disposal.

➢ Advantages of Eco-Friendly Production

✓ Efficiency and savings

While sustainable practices may require a larger outlay of capital up front, they typically result in substantial savings for manufacturers over time. Reduced operational costs are a result of energy efficiency, less waste, and optimized supply chains. Manufacturers can save even more money in the long run by extending the life of their equipment and raw materials through more efficient resource utilization.

✓ Improved customer loyalty and brand image

Customers are placing a higher value on sustainable products and services, giving green businesses a competitive advantage in the market. Businesses can increase their customer loyalty among environmentally concerned consumers by highlighting their commitment to responsible business practices. This promotes long-term customer relationships while also improving the brand's reputation.

✓ Minimizing risk and ensuring compliance with regulations

The environmental regulations are getting tougher all over the globe, and companies that prioritize sustainability will be better able to keep up. In order to stay out of legal hot water and reduce their carbon footprint, businesses must follow all applicable environmental regulations.

✓ Leading the industry in innovation

Sustainability drives the development of new products, production methods, and business models. Many people look up to companies that are pioneers in green manufacturing because they are at the forefront of industry innovation and can seize new opportunities in sustainable products and green technology.

➢ Lean Six Sigma contributes to the sustainability of businesses

✓ Minimizing environmental waste through lean methodologies

- Lean methodologies seek to reduce waste and enhance value, thereby fostering environmental stewardship.
- Lean methodologies intricately connect the concept of 'Reduce, Reuse, Recycle' with the objective of eradicating waste in every conceivable form.
- The automotive sector has employed lean methodologies to optimise material utilisation and enhance production efficiency, thereby diminishing hazardous waste and emissions.
- Lean methodologies advocate for the judicious utilization of resources, including raw materials and energy, thereby fostering a culture of conservation and a heightened awareness of environmental stewardship.

✓ The application of six-sigma principles can improve energy efficiency.

- The Six Sigma methodology emphasizes the importance of data analysis and process enhancement and aims to identify and eradicate inefficiencies that contribute to excessive energy consumption.
- The application of data analysis and process enhancement has enabled organizations to markedly diminish their energy consumption footprints.
- In data centers, Six Sigma initiatives have refined cooling systems and enhanced server efficiency, resulting in significant energy savings while maintaining service quality.

✓ Optimisation of Sustainable Supply Chains

- Lean Six Sigma methodologies provide a structured approach to enhancing the efficiency and sustainability of supply chains.
- Organizations can effectively reduce overstock and curtail inventory management waste through the application of Lean principles, thereby reducing the carbon footprint associated with storage and spoilage.
- Empirical evidence-based methods enable the optimization of logistics and transportation routes, resulting in reduced fuel consumption and reduced emissions.

✓ Fostering an Environment of Ongoing Advancement for Sustainable Practices

- The principles of LSS, cantered around the pursuit of ongoing enhancement, resonate deeply with sustainability's objectives.
- This philosophy advocates for organizations to engage in an ongoing evaluation and enhancement of their environmental performance.

✓ We are utilising technological advancements to enhance sustainable process optimisation.

- Lean Six Sigma methodologies leverage the capabilities of advanced analytics, IoT, and AI to provide real-time insights into process performance.
- Through the amalgamation of these technologies with LSS principles, organizations can achieve a degree of precision in their sustainability initiatives that was once beyond reach.

4. Contributions and future agenda

This research offers a fresh perspective on how Green Lean Six Sigma (GLSS) relates to current technological advances. The research highlights how digital advances—with a particular focus on GLSS—can improve sustainable development strategies. This study builds on earlier work that used Lean Six Sigma techniques to establish frameworks for sustainable innovation; however, it highlights the possibility of operational excellence by integrating contemporary innovations into GLSS. The results show that GLSS and modern technology can work together to strengthen sustainability objectives and boost operational performance. However, the study acknowledges several limitations, including its short duration, which hinders the ability to draw definitive conclusions about the impact over time and the significant variation in technology adoption levels across different businesses. We highly recommend putting more research into longitudinal studies and comprehensive examinations to further understand the broad implications and long-term consequences of these technologies on social responsibility and environmental sustainability. Research concludes that state-of-the-art technologies are crucial to making GLSS approaches more effective and that companies can reap significant operational and environmental benefits by carefully incorporating these technologies while continuously improving their processes. Future research should investigate the impacts of GLSS and advanced technologies on organizational performance and other dimensions of sustainability. It could also examine the linkages between the circular economy, GLSS, and digital innovation with practical implications. Additional case studies and qualitative research are required to validate these findings. The study recommends further research to enhance the model for replication in various industry contexts.

5. Conclusions

Viewed from a different angle, the GLSS paradigm signifies a forward-thinking change in production methods, especially within the advanced technologies. In addition to promoting environmental friendly products, this strategy highlights the importance of organizational learning and system improvements, which help alleviate environmental concerns. By incorporating GLSS into their processes, manufacturers can contribute to the creation of a circular economy, ensuring efficient resource use in accordance with the three R's: reduce, reuse, and recycle. Sustainability drives the development of new products, production methods, and business models. Many people look up to companies that are pioneers in green manufacturing because they are at the forefront of industry innovation and can seize new opportunities in sustainable products and green technology. If a company wants to remain competitive and secure its operations for the future, it must implement a strategy that prioritizes sustainability in manufacturing. Manufacturers can lessen their impact on the environment and increase productivity and creativity by implementing energy-saving measures, following the principles of the circular economy, and purchasing state-of-the-art green technology. Manufacturers who put an emphasis on environmental friendly practices will benefit the world in two ways: first, by meeting the increasing demand for sustainable products, and second, by securing their own financial future. A more environmental friendly and efficient future is within reach, and sustainable manufacturing is a critical component. Another perspective is to view sustainability initiatives through the lens of Six Sigma, a transformative method that reimagines corporate responsibility while simultaneously enhancing operational efficiency. Long associated with quality improvement, Six Sigma's recent applications in environmental stewardship show how companies are starting to prioritize a more comprehensive definition of success that considers environmental impact. More and more, businesses that are embracing Six Sigma principles are realizing that reducing their impact on the environment is just as important as increasing their profit margins. By bridging the gap between operational excellence and sustainable practices, this methodology proves that being efficient and environmentally conscious are not mutually exclusive. Businesses can save money and help the environment by using data analytics to find inefficient ways of doing things and then finding better ways to use their resources. The Six Sigma focus on energy efficiency also demonstrates a commitment to combating climate change. To set themselves apart as sustainability leaders, companies aren't merely complying with regulations; they're actively looking for ways to innovate while reducing energy consumption. This innovative strategy not only boosts the company's image, but it also appeals to eco-conscious buyers. The emphasis on a sustainable product life cycle also shows a serious desire to do the right thing by the environment. Businesses can make sustainable, financially rewarding products by thinking about the environment at every step of the product life cycle. Recognizing the interconnectedness between long-term success and environmental health, there has been a shift toward circular practices. Essentially, when seen through the perspective of sustainability, Six Sigma becomes apparent as a powerful tool for transformation in contemporary business. In the end, it helps the economy and the environment by encouraging a culture of green excellence and enabling organisations to accomplish operational goals while also embracing a larger mission of environmental stewardship. Furthermore, a comprehensive approach to sustainability is required due to the interaction between green and LSS practices.

Both methods have their benefits, but using only one may make it challenging to balance economic, environmental, and social factors. Synergies between lean and green practices can improve sustainability performance. Although combining these practices isn't without its difficulties, it shows that when done right, it can improve results. We suggest ways to make the GLSS model more widely used by weighing the advantages and disadvantages of this integration. The ultimate goal of this combined approach, which is considered a critical evolution in manufacturing, is to harmonize the three sustainability pillars and create a more resilient and responsible industrial landscape.

- Improving the efficiency of green, lean, Six Sigma, and other new technologies in the industry greatly impacts the sustainability of corporations.
- Smart automation, real-time monitoring, and analytics that can foresee the future all contribute to more sustainable operations.
- Such innovations enhance operational efficiency and environmental friendliness, thereby supporting Green Lean Six Sigma initiatives.
- The analysis shows that an increasing number of companies are settling on environmentally conscious policies and procedures.
- Obstacles to using new technologies include opposition to change, high implementation costs, and lack of technological knowledge.
- To effectively incorporate new technologies and get beyond these challenges, we counsel a tailored support plan.
- Sustainable application of new technologies and GLSS concepts usually depends on cooperation among departments, suppliers, customers, and industrial partners.
- Merging these technologies with GLSS methods could lead to a new benchmark for operational success in the long run.

Contributing to what is already known, this chapter shows how GLSS and new technologies can collaborate to make sustainability better. The limited time frame precludes drawing any firm conclusions about the new technologies' potential long-term impacts or tracking the development of operational and environmental performance over the years.

References

Abbasi, S. and Choukolaei, H. A. (2023). A systematic review of green supply chain network design literature focusing on carbon policy. Decision Analytics Journal 6(1–16): 100189. https://doi.org/10.1016/j.dajour. 2023.100189.

Abualfaraa, W., Salonitis, K., Al-Ashaab, A. and Ala'raj, M. (2020). Lean-Green Manufacturing Practices and Their Link to Sustainability'. https://doi.org/10.3390/su12030981.

Akbar, M. and Irohara, T. (2018). Scheduling for sustainable manufacturing: A review. Journal of Cleaner Production 205: 866–883. https://doi.org/ 10.1016/j.jclepro.2018.09.100.

Alexa, L., Pîslaru, M. and Avasilcăi, S. (2022). From Industry 4.0 to Industry 5.0—An Overview of European Union Enterprises, Springer, p. 221–231.

Anass, C., Amine, B., Ibtissam, E. H., Bouhaddou, I. and Elfezazi, S. (2021). Industry 4.0 and lean six sigma: Results from a pilot study. pp. 613–619. In Advances in Integrated Design and Production:

Proceedings of the 11th International Conference on Integrated Design and Production, CPI 2019, October 14–16, 2019, fez, Morocco. Springer International Publishing.
Antonioli, D., Ghisetti, C., Mazzanti, M. and Nicolli, F. (2022). Sustainable production: The economic returns of circular economy practices. Business Strategy and the Environment 31(5): 2603–2617. https://doi.org/ 10.1002/bse.3046.
Bai, C., Dallasega, P., Orzes, G. and Sarkis, J. (2020). Industry 4.0 technolo gies assessment: A sustainability perspective. International Journal of Production Economics 229: 107776. https://doi.org/10.1016/j.ijpe. 2020.107776.
Belhadi, A., Kamble, S. S., Zkik, K., Cherrafi, A. and Touriki, F. E. (2020). The integrated effect of Big Data Analytics, Lean Six Sigma and Green Manufacturing on the environmental performance of manufacturing companies: The case of North Africa. Journal of Cleaner Production 252: 119903. https://doi.org/10.1016/j.jclepro.2019.119903.
Ben Ruben, R., Vinodh, S. and Asokan, P. (2018). Lean Six Sigma with environmental focus: review and framework. The International Journal of Advanced Manufacturing Technology. https://doi.org/10.1007/s00170-017-1148-6.
Boumsisse, I. and Benhadou, A. H. M. (2024). Exploring the potential synergies between Industry 5.0 and Green Lean Six Sigma for sustainable performance: a new dimension of operational excellence. DOI: 10.70082/esiculture.vi.1442.
Buer, S.-V., Strandhagen, J. O. and Chan, F. T. S. (2018). The link between industry 4.0 and lean manufacturing: Mapping current research and establishing a research agenda. International Journal of Production Research 56(8): 2924–2940. https://doi.org/10.1080/00207543. 2018.1442945.
Ciano, M. P., Dallasega, P., Orzes, G. and Rossi, T. (2020). One-to-one relationships between Industry 4.0 technologies and Lean Production techniques: A multiple case study. International Journal of Production Research. https://doi.org/10.1080/ 00207543.2020.1821119.
Ciliberto, C., Szopik-Depczynska, K., Tarczynska-Łuniewska, M., Ruggieri, A. and Ioppolo, G. (2021). Enabling the circular economy transition: A sustainable lean manufacturing recipe for industry 4.0. Business Strategy and the Environment 30(7): 3255–3272. https://doi.org/10.1002/bse.2801.
Cherrafi, A., Chiarini, A., Belhadi, A., El Baz, J. and Chaouni Benabdellah, A. (2022). Digital technologies and circular economy practices: Vital enablers to support sustainable and resilient supply chain management in the post-COVID-19 era. The TQM Journal 34(7): 179–202. https:// doi.org/10.1108/TQM-12-2021-0374.
Cherrafi, A., Elfezazi, E., Hurley, B., Garza-Reyes, J. A., Vikas Kumar, V., Anosike, A. et al. (2019). Green and lean: a Gemba–Kaizen model for sustainability enhancement. https://doi.org/10.1080/09537287.2018.1501808.
Cherrafi, A., Elfezazi, S., Govindan, K., Garza-Reyes, J. A., Benhida, K. and Mukhlis, A. (2017). A framework for the integration of green and lean six sigma for superior sustainability performance. International Journal of Production Research 55(15): 4481–4515. https://doi.org/10.1080/00207543.2016.1266406.
Coca, L-C. and Pîslaru, M. (2022). Innovation and productivity growth in manufacturing industry-pharmaceutical company study case. The 18th Romanian Textiles and Leather Conference. DOI: https://doi.org/10.2478/9788367405133-055.
Di Maria, E., De Marchi, V. and Galeazzo, A. (2022). Industry 4.0 technologies and circular economy: The mediating role of supply chain integration. Business Strategy and the Environment 31(2): 619–632. https:// doi.org/10.1002/bse.2940.
Dües, C. M., Tan, K. H. and Lim, M. (2013). Green as the new lean: how to use lean practices as a catalyst to greening your supply chain. Journal of Cleaner Production 40: 93–100. https://doi.org/10.1016/j.jclepro.2011.12.023.
Edwin Cheng, T. C., Kamble, S. S., Belhadi, A., Ndubisi, N. O., Lai, K. and Kharat, M. G. (2022). Linkages between big data analytics, circular economy, sustainable supply chain flexibility, and sustainable performance in manufacturing firms. International Journal of Production Research 60(22): 6908–6922. https://doi.org/10.1080/00207543. 2021.1906971.
Fayyaz, A., Liu, C., Xu, Y. and Ramzan, S. (2024). Effects of green human resource management, internal environmental management and developmental culture between lean six sigma and operational performance. International Journal of Lean Six Sigma. https://doi.org/10.1108/IJLSS-04-2023-0065.

Garza-Reyes, J. A. (2015). Lean and green—a systematic review of the state of the art literature. Journal of Cleaner Production 102: 18–29. https://doi.org/10.1016/j.jclepro.2015.04.064.

Garza-Reyes, J. A., Kumar, V., Chaikittisilp, S. and Tan, K. H. (2018). The effect of lean methods and tools on the environmental performance of manufacturing organisations. International Journal of Production Economics 200: 170–180. https://doi.org/10.1016/j.ijpe.2018.03.030.

Gatell, I. S. and Avella, L. (2023). Impact of Industry 4.0 and circular economy on lean culture and leadership: Assessing digital green lean as a new concept. Research on Management and Business Economics. https://doi.org/10.1016/j.iedeen.2023.10023.

Ghaithan, A., Khan, M., Mohammed, A. and Hadidi, L. (2021). Impact of Industry 4.0 and lean manufacturing on the sustainability performance of plastic and petrochemical organizations in Saudi Arabia. Sustainability 13(20): 11252. https://doi.org/10.3390/su132011252.

Hussain, K., He, Z., Ahmad, N., Iqbal, M. and Taskheer Mumtaz, S. M. (2019). Green, lean, six sigma barriers at a glance: A case from the construction sector of Pakistan. Building and Environment Building and Environment. DOI: 10.1016/j.buildenv.2019.106225.

Kamble, S., Gunasekaran, A. and Dhone, N. C. (2020). Industry 4.0 and lean manufacturing practices for sustainable organisational performance in Indian manufacturing companies. International Journal of Production Research 58(5): 1319–1337. https://doi.org/10.1080/00207543. 2019.1630772.

Khan, M., Ajmal, M. M., Jabeen, F., Talwar, S. and Dhir, A. (2023). Green sup ply chain management in manufacturing firms: A resource-based view point. Business Strategy and the Environment 32(4): 1603–1618. https://doi.org/10.1002/bse.3207.

Kaswan, M. S. and Rathi, R. (2020). Green lean six sigma for sustainable development: Integration and framework. Environmental Impact Assessment Review. DOI: 10.1016/j.eiar.2020.106396.

Kaswan, M. S., Rathi, R., Garza -Reyes, J. A. and Antony, J. (2023). Exploration and investigation of green lean six sigma adoption barriers for manufacturing sustainability. IEEE Transactions on Engineering Management. DOI: 10.1109/TEM.2021.3108171.

Kumar, S., Luthra, S., Govindan, K., Kumar, N. and Haleem, A. (2016). Barriers in Green Lean Six Sigma product development process: An ISM approach. Prod Plan Control 27(7-8): 604–620.

Letchumanan, L. T., Gholami, H., Yusof, N. M., Ngadiman, N. H. A. B., Salameh, A. A., Štreimikienė, D. et al. (2022). Analyzing the factors enabling green lean six sigma implementation in the Industry 4.0 Era. Sustainability 14(6): 3450. https://doi.org/10.3390/su14063450.

Lu, H., Zhao, G. and Liu, S. (2022). Integrating circular economy and industry 4.0 for sustainable supply chain management: A dynamic capability view. Production Planning and Control 1–17. https://doi.org/10.1080/ 09537287.2022.2063198.

Ma, Q., Murshed, M. and Khan, Z. (2021). The Nexuses between Energy Investments, Technological Innovations, Emission Taxes, and Carbon Emissions in China. https://doi.org/10.1016/j.enpol.2021.112345.

Mathiyazhagan, K., Sengupta, S. and Mathivathanan, D. (2019). Challenges for implementing green concept in sustainable manufacturing: A systematic review. Opsearch 56: 32–72. https://doi.org/10.1007/ s12597-019-00359-2.

Miraute, L.-C. and Pîslaru, M. (2024). Challenges of Sustainability Assessment in the Manufacturing Industry. International Symposium "Technical Textiles - Present and Future". Edition 2023. DOI: https://doi.org/10.2478/9788367405355-040.

Mittal, V. K., Sindhwani, R. and Kapur, P. K. (2016). Two-way assessment of barriers to lean–green manufacturing system: Insights from India. International Journal of Systems Assurance Engineering and Management. DOI: 10.1007/s13198-016-0461-z.

Neligan, A., Baumgartner, R. J., Geissdoerfer, M. and Schöggl, J. P. (2023). Circular disruption: Digitalisation as a driver of circular economy business models. Business Strategy and the Environment 32(3): 1175–1188. https://doi.org/10.1002/bse.3100.

Pandey, H., Garg, D. and Luthra, S. (2018). Identification and ranking of enablers of green lean six sigma implementation using AHP. International Journal of Productivity and Quality Management. DOI: 10.1504/IJPQM.2018.089156.

Romero, D. and Flores, M. (2019). Five management pillars for digital transformation integrating the lean thinking philosophy. In Proceedings 2019 IEEE International Conference on Engineering, Technology, and Innovation, ICE/ITMC 2019. https://doi. org/10.1109/ICE.2019.8792650.

Rosa, P., Sassanelli, C., Urbinati, A., Chiaroni, D. and Terzi, S. (2020). Assessing relations between circular economy and industry 4.0: A systematic literature review. International Journal of Production Research 58(6): 1662–1687. 1680896. https://doi.org/10.1080/00207543.2019.

Rosin, F., Forget, P., Lamouri, S. and Pellerin, R. (2020). Impacts of Industry 4.0 technologies on Lean principles. International Journal of Production Research 58(6): 1644–1661. https://doi.org/10.1080/00207543.2019.1672902.

Ruben, R. B., Vinodh, S. and Asokan, P. (2018). Lean Six Sigma with environmental focus: review and framework. Int. J. Adv. Manuf. Technol. 94: 4023–4037. https://doi.org/10.1007/s00170-017-1148-6.

Saabye, H., Kristensen, T. B. and Wæhrens, B. V. (2020). Real-time data utilization barriers to improving production performance: An in-depth case study linking lean management and industry 4.0 from a learning organization perspective. Sustainability 12(21): 1–21. https://doi.org/10.3390/su12218757.

Siegel, R., Antony, J., Garza-Reyes, J. A., Cherrafi, A. and Lameijer, B. (2029). Integrated green lean approach and sustainability for SMEs: From literature review to a conceptual framework. Journal of Cleaner Production. https://doi.org/10.1016/j.jclepro.2019.118205.

Shamsi, M. A. and Alam, A. (2018). Exploring lean six sigma implementation barriers in information technology industry. International Journal of Lean Six Sigma 9(4): 523–542. https://doi.org/10.1108/IJLSS-06-2017-0054.

Skalli, D., Charkaoui, A., Cherrafi, A., Garza-Reyes, J. A., Antony, J. and Shokri, A. (2023). Analyzing the integrated effect of circular economy, Lean Six Sigma, and Industry 4.0 on sustainable manufacturing performance from a practice-based view perspective. https://doi.org/10.1002/bse.3546.

Skalli, D., Charkaoui, A. and Cherrafi, A. (2022a). The integration of industry4.0 in operations management: Toward smart lean six sigma. pp. 3–11. In International Conference on Digital Technologies and Applications. Springer. https://doi.org/10.1007/978-3-031-01942-5_1.

Skalli, D., Charkaoui, A., Cherrafi, A., Garza-Reyes, J. A., Antony, J. and Shokri, A. (2023). Industry 4.0 and lean six sigma integration in manufacturing: A literature review, an integrated framework and proposed research perspectives. Quality Management Journal 30: 16–40. https://doi.org/10.1080/10686967.2022.2144784.

Suresh, P., Dillibabu, S. P., Kumar, B. A., Paul, A., Boopathi, S. and Ramalakshmi, D. (2024). Strategies for Carbon Footprint Reduction in Advancing Sustainability in Manufacturing (pp. 317–350). IGI Global. https://doi.org/10.4018/979-8-3693-3625-0.ch012.

Thiruvarasu Letchumanan, L., Gholami, H., Yusof, N. M., Bin Ngadiman, N. H. A., Salameh, A. A., Dalia Štreimikienė, D. et al. (2022). Analyzing the Factors Enabling Green Lean Six Sigma Implementation in the Industry 4.0 Era. https://doi.org/10.3390/su14063450.

Tortorella, G. L., Giglio, R. and van Dun, D. H. (2019). Industry 4.0 adoption as a moderator of the impact of lean production practices on operational performance improvement. International Journal of Operations and Production Management 39: 860–886. https://doi.org/10.1108/IJOPM-01-2019-0005.

Yacob, P., Wong, L. S. and Khor, S. C. (2019). An empirical investigation of green initiatives and environmental sustainability for manufacturing SMEs. Journal of Manufacturing Technology Management. doi:10.1108/JMTM-08-2017-0153.

Index

For Product Safety Concerns and Information please contact our EU representative GPSR@taylorandfrancis.com
Taylor & Francis Verlag GmbH, Kaufingerstraße 24, 80331 München, Germany

www.ingramcontent.com/pod-product-compliance
Ingram Content Group UK Ltd.
Pitfield, Milton Keynes, MK11 3LW, UK
UKHW022211190926
475702UK00003B/300

9781032840123